PIPELINE INTEGRITY HANDBOOK

PIPELINE INTEGRITY HANDBOOK

Risk Management and Evaluation

RAMESH SINGH

AMSTERDAM • BOSTON • HEIDELBERG • LONDON
NEW YORK • OXFORD • PARIS • SAN DIEGO
SAN FRANCISCO • SINGAPORE • SYDNEY • TOKYO
Gulf Professional Publishing is an imprint of Elsevier

Gulf Professional Publishing is an imprint of Elsevier
225 Wyman Street, Waltham, MA 02451, USA
The Boulevard, Langford Lane, Kidlington, Oxford, OX5 1GB, UK

First edition 2014

British Library Cataloguing in Publication Data
A catalogue record for this book is available from the British Library

Library of Congress Cataloging-in-Publication Data
A catalog record for this book is available from the Library of Congress

ISBN: 978-0-12-387825-0

For information on all Gulf Professional Publishing visit our web site at books.elsevier.com

Printed and bound in United States of America

13 14 15 16 17 10 9 8 7 6 5 4 3 2 1

DEDICATION

This book is dedicated to the memory of my late father, Kripashanker Singh, from whom I learned to put pen to paper, who taught me to be analytical in my approach, and who instilled in me a sense of discipline and dedication to the work I do.

CONTENTS

PREFACE

There are several information sources on the market on this subject; most of them are regulatory requirements and guidelines. In most cases, due their very nature, the regulatory materials are specifically tailored as 'do and don't' lists. This rule of thumb hand book is intended to bring closer the management principles and regulatory requirements and to apply these principles to achieve the best possible results.

While the regulatory requirements are essential, the application of management principles, in the writer's view, uplifts the degree of success of the integrity management efforts of a company. This intends to support the under-supported by giving a practical perspective to the theoretical texts.

The book is aimed at those managers, engineers, and non-engineers, who are responsible for establishing and managing pipeline integrity for public safety. The book is intended to serve as a body of knowledge and as a source of reference.

In writing this book I do not claim originality on all thoughts and words, as this would be impossible on a subject as universal as integrity management. I acknowledge various sources and job positions that have contributed to my experience of the subject, and I am proud of them. Where I have consciously borrowed matters and ideas directly from these experiences and resources, I have acknowledged them as best as I can.

Those individuals who need more detailed information on any specific topic covered in this book should reach out to these acknowledged specialized associations, institutions, and local regulatory bodies for further guidance. There are several published works available from these bodies that can be of immense help in developing in-depth understanding of specific subjects.

ACKNOWLEDGMENT

Writing this book made me realize how dependent a person is in accomplishing a task of this nature. The process started with retrieving several years of notes, handouts, and hand written chits.

I am extremely grateful to the management and team of Gulf Interstate Engineering, Houston (www.gie.com), for creating an environment that encouraged me to write this book. I am also grateful to the clients of Nadoi Management, Inc. who allowed me to use data and pictures from the work that I performed for them.

I am also indebted to the encouragement, support, and help from my friend Olga Ostrovsky. She helped me negotiate the obstacles of writing and editing the drafts. Without her expert help, this book would not have been possible.

Last but not least, I am also grateful to my loving wife, Mithilesh, and my son, Sitanshu, for their support in helping me to accomplish this goal. Mithilesh tolerated my indulgence with the project. Without her support and understanding, this task would not have been possible.

ABOUT THE AUTHOR

Ramesh Singh, MS, IEng, MWeldI, is registered as an Incorporated Engineer with the UK British Engineering Council. He is also a member of The UK Welding Institute. He is a Senior Principal Engineer (Materials, Welding, and Corrosion) for Gulf Interstate Engineering, Houston, TX, USA.

Ramesh is a graduate of the Indian Air force Technical Academy, with diplomas in Structural Fabrication Engineering and Welding Technology. He has been a member and officer of the Canadian Standard Association and NACE and has served on several technical committees. He has worked in industries spanning aeronautical, alloy steel castings, fabrication, machining, welding engineering, petrochemical, and the oil and gas industries. He has written several technical papers and published articles in leading industry magazines, addressing the practical aspects of welding, construction, and corrosion issues relating to structures, equipments, and pipelines.

He provides management consultancy in his areas of expertise through Nadoi Management, Inc.

PART ONE

Pipeline Risk Management and Engineering Assessment

CHAPTER ONE

Introduction

Pipeline Integrity Handbook
ISBN 978-0-12-387825-0

Risk management (RM) has been embraced by both the pipeline industry and regulatory agencies as a way to increase public safety and also to optimize all aspects of pipeline design, operations, and maintenance. The focus of RM is to establish a program that follows industry best practices, gives the pipeline owner and operator a long-term decision support tool, and instills confidence in the public about the safe operations of pipelines passing through their neighborhood.

The RM process typically begins with a review of the risks associated with the specific pipeline systems, compares it with the risk management concepts and methodologies, and then focuses on the most effective risk management techniques that can be applied. These techniques are currently in use by the pipeline industry in this process. The estimate is made of the severity of pipeline releases in terms of:

- The potential volume of product that could be released
- The physical pathways and dispersion mechanisms by which the product could move to a high consequence area (HCA)

Table 1-1-1 Mandatory Assessment of Integrity (ASME B 31.8S).

Pipe operating above percentage SMYS	First inspection after construction within (years)
Above 60	10
Above 50 but less than 60	13
Above 30 but less than 50	15
Below 30	20

ASME B31.8S gives a list of criteria presented in Appendix A that address subjects such as pipeline material, design conditions, construction, and inspection and operating history. It sets out some strict guidelines for inspection frequency based on the percentage of specified minimum yield strength (SMYS) over the years.

- The amount of product that might actually reach the boundaries of the HCA and
- The population and environmental resources that could be affected by such a release.

The emphasis throughout is on practical, ready-to-apply techniques that would yield positive and cost effective benefits.

The risk management process can be structured so that it is appropriate for application either to a new or an existing pipeline system. Understanding the concepts and principles listed below helps the risk manager to focus on the task as a more knowledgeable person:

- Basic concepts of risk
- Risk assessment processes
- The indexing technique
- Failure modes
- Consequence analysis
- Hazard zone calculations
- Leak impact factor
- Supplemental assessments
- Data collection and analysis
- QA/QC of data
- Dynamic segmentation
- Using common spreadsheet and desktop database tools
- Managing the risks
- Resource allocation modeling
- Practical applications
- Integrity management and risk management.

Subsequent chapters of this part of the book will delve into some of the following concepts and principles:

- Concept of risk management and risk defined
- Data collection and analysis
- Risk assessment concepts and tools
- Identification of hazards that lead to failure
- Determining consequences of failure and identification of HCA.

CHAPTER TWO

Basic Concepts of Risk Management and Risk Defined

Contents

Pipeline Integrity Handbook
ISBN 978-0-12-387825-0

Definition of failure

- **An unintentional release of the pipeline's product, or loss of integrity**
- **Failure to perform its intended function**
- **Examples include leaks in pipelines due to internal corrosion, external corrosion, improper operation, or third party damage**

WHAT IS RISK?

Before one starts on the process of managing risk, one needs to know what risk is, and furthermore be able to identify risk. The most common definition of risk is the relationship between the probability of an incident's occurrence and the consequence of that occurrence. This can be written as follows:

$$R = P \times C$$

where R = risk, P = probability, and C = consequence

If we observe the above equation we note that there are only two components to risk, and if we take one of these out of the equation, the entire risk will be eliminated. This is important, as the primary goal of risk management is to eliminate or contain risk:

1. The probability of failure and
2. The consequences of failure.

Definition of probability

- **How likely is it that the risk will to occur? (What are the chances of failure?)**
- **Examples include the degree of belief that an event will occur based upon assessment of risk of failure. Crossing a fault zone increases the probability of failure.**

Definition of consequenses

- **The results of a failure**
- **Example: A pipeline failure close in proximity to areas of high density population or near a school, hospital, or any other public sites can cause an explosion and fire with a considerable loss of life and property.**

RELATIONSHIP BETWEEN RISK AND INTEGRITY MANAGEMENT

The integrity management (IM) concept is based on control and elimination of risk by assessing probability of failure. As stated above, the probability and consequence relationship establishes the level of risk. In other words the IM is really a way to address the probability of failure which may pose a threat to a high consequence area (HCA). Any condition that poses a threat to the integrity also increases the probability of failure and, hence, it is a risk. The consequences of failure could be low or high. When the probability of failure is high in a component that can have higher consequences in terms of loss and damage to lives or property, it is termed an HCA component.

A failure in HCA is likely to cause more damage to life and property. Its significance will be more pronounced, and hence it assumes higher risk. As a result HCAs assume highest priority in the application of IM principles. One of the primary steps in IM application is to identify and recognize the HCA in a system. Such proactive steps taken toward the mitigation of these consequences of failures are called risk mitigation. The relationship between risk mitigation, IM, and HCA is expressed as:

$$\text{Risk Mitigation} = \text{IM} \times \text{HCA}$$

Definition of risk management

- **The reaction to perceived risks.**
- **Example: A strategy that mitigates risk for a specific area. If pipeline coating has failed – fix it!**

What is risk assessment?

An important part of this process is the risk assessment. There is no universally accepted way to assess risks from a pipeline.

Risk assessment programs could be either performance based or prescriptive based.

Definition of risk assessment

- **Risk assessment is a measuring process and a risk model is a measuring tool**

Risk = Probability x Consequences
(R = P x C)

The perspective-based program is a tool that complements the IM program by organizing data and being helpful in integrity management decision making.

The performance-based program addresses the following objectives:

1. It organizes the data and prioritizes the plan of action.
2. It decides on the timing of and selection of inspection method and prevention or mitigation plan.

It is important to identify the limits and possibilities of any risk assessment process. The RM process is not a crystal ball, whereby one can see and forecast the location and time of pipeline failure. Most pipeline accidents are the result of several system failures on the part of operators and what the risk assessment system does is monitor the effective functioning of the system. At best the available risk assessment methodologies provide an indirect way to predict a probability of failure; this, however, is subject to the accuracy of the data input in to the prediction model.

Assessment is in fact an effort to systematically and objectively capture everything that can be known about the pipelines and their environments. All of this collected data on the risk context is used and applied to determine the probability of failure on a generally established scale, allowing for an informed decision. The model developed by using the data should be comprehensive enough, because it can process more information than a single person can. The effectiveness of a risk assessment model can be maximized by following the steps discussed below.

The assessment process should be set up with a clear definition of objectives to be achieved. The objective of a risk assessment program could be any or a combination of the following:

- To assess the effect of mitigation action already taken
- To determine the most effective mitigation step for a known threat
- To assess the impact of change in the inspection schedule
- To prioritize sections of a pipeline system for integrity assessment or mitigation action
- To make changes in the inspection method or to arrange reallocation of resources.

The models are developed on relative assessment, scenario-based or probabilistic approaches. Whatever the approach selected in building a model it must be checked, rechecked, and validated. Lots of questions must be asked to establish that the result obtained is capable of addressing specific objectives of the operator RM program.

Test of acquired knowledge

The risk model should be able to do more than a single person or the combined brains of a few consultants. It is not humanly possible to take cognizance of several variables and data and analyze them to reach an effective decision.

The model developed for the risk assessment should be able to simultaneously take note of hundreds of thousands of variables hidden in the collected data.

The model should be able to bring to the table information that was not previously known and it should be able to present some surprising new information. Such surprises should be revalidated by further research and data analysis. The net result of all of this should be the re-evaluation of the current integrity model and identifying ways to make improvements.

Room for complacency – results not to be taken for granted

The true scientific approach to any issue is to ask questions. Any surprising new information presented as a result of the data analysis must make the operator ask, "Why?"

Why is this section of the system a high risk?
Why this new information now?
Why was it not known before?
Question the premise. Is that premise correct?

Being skeptical of new knowledge is the path to validation of that new knowledge. The model should be able to respond to these questions. It may be able to give such reasons as:

- There is a new senior citizen home in the area.
- The school district has opened a new school in the proximity.
- There is an increase in population density.
- A vulnerable aquifer is found that was not known before.
- A new spur of highway passes through the lease.
- In line inspection (ILI) has not been carried out for several years.
- Several coating failures have been reported for the system.

Such validation of new knowledge would make the model creditable and acceptable.

Know the pipe system and associated risk

Through the risk assessment process about any segment of pipeline system, the integrity manager should be able to find out specifically the corrosion risk, the third party risk, the types of receptors, and the spill volume.

In one instance, the company Pipeline Integrity Manager and Chief Inspector had used a practical way to mark on each field map the following data and post them on the company website for everyone to view. The failures were graded and color-coded for level of risk and easy recognition:

- Volume of leak flow
- Number of days the system was shut down for repair
- Production loss

- Human and environmental damage caused
- Evacuation of native population from around the gas leak area and
- Total cost to the company.

This appears to be a very obvious thing for a risk assessment to do, but surprisingly, not everyone seems to know or, better still, practices the obvious. In another operating company, some field supervisors had their own notion of risk and they had assigned risk levels to their respective areas. Others did not retain information specific to a given location. They were essentially satisfied with periodic submission of collected data to local regulatory authorities. It took some serious efforts to change the mindset of the field supervisors through meetings, trainings, and reviews to bring everyone on board with a common understanding to make the program a success.

Measure the completeness of the assessment model

Questions should be asked about possible threats. Questions such as, what about the native reserve across the plant? What if the leak occurs near a river? How about SCC possibilities? There have been MIC cases in other pipelines in the area concerning the level of MIC risk to this pipeline.

All the probability issues must be identified and addressed. All known failure modes should be considered, even if they are very rare for the specific system.

Very complex consequence potential should be assessed in a way that would with stand the need of the system for a long time and cover most ground. All receptors, sensitivities, and variables must be addressed. A complete consequence evaluation will consider at least these four variables:

1. Spill sizes
2. Leak detection
3. Emergency response (receptors)
4. Product characteristics.

Thus the Consequence = Spill × Spill Size (spread) × Receptors × Product Characteristics (Hazard).

If the numerical value of any of these goes to zero, then there are no consequences, no matter how bad the other three are.

Relative risk versus absolute risk

Both relative risk and absolute risk approaches have their advantages. It is up to the Risk Manager to take the best of both. In that respect, steps should be initiated to use relative scores for routine day-to-day management, while keeping the option for switching over to the absolute model for long-term planning, if and when it is considered necessary.

The models indicate that the relationship between an absolute failure probability scale and a relative scale is defined by some curve that is asymptotic to at least one axis, either beginning flat or beginning steep.

If a good scoring model is developed it should be able to show that at one end of the spectrum is a pipeline without any safety provisions. It is operated in the most hostile environment and a failure is imminent.

At the other end of the spectrum is a system which can be termed as a "bulletproof" version. It is buried 20 ft deep, it has double × heavy wall, and the material is crack resistant and corrosion resistant. The system also has secondary containment, the ROW is fenced and guarded around the clock, and a team of technicians regularly monitors the integrity by inspection and verification. One may say the system is failure proof, however utopian this might appear. These extreme positions at either end of the spectrum are very well understood. It is the middle regions that are to be understood and cared for. The middle region is the most critical, and that is where additional data will be required to finalize the curve.

It should be noted that there is the possibility of misjudging a variable. In spite of the quantification, some risk factors may be less than perfect. The results may present a reliable picture of sections that have relatively fewer adverse factors along with those that have relatively more adverse factors in the analysis.

The risk assessment therefore involves recognition and identification of threats to a pipeline system and then initiates the mitigating steps. The system involves preemptive and proactive recognition and action to prevent accidents as compared to post-failure salvaging and corrective measures. All this is possible if accurate and exhaustive data is gathered and analyzed.

CHAPTER THREE

Data Collection

Contents

Pipeline Integrity Handbook
ISBN 978-0-12-387825-0

Pipeline integrity elements

- **The following are crucial in a PIM**
  - **Location data**
  - **Operational data**
  - **Original design data**
  - **Pigging data**
  - **Chemical program data**
  - **Cathodic protection data**
  - **Coating data**
  - **Monitoring & inspection data**

ROLE OF DATA COLLECTION

For a successful risk assessment and integrity management program it is important to understand the critical role that the collection of data plays. A diligent effort must be made to collect all possible data relating to the section of pipeline that needs to be assessed and managed. There is no shortcut to the collection process. The computerization of data storage has reduced the legwork; however, it has also increased the size of available data files. Greater quantities of data make better analysis, as evaluation of large data can reveal details that otherwise may not be visualized.

The collection of data is the primary step to analysis. In the subsequent sections the type of data that must be collected for analysis is discussed. These data are pipe-segment specific and in the USA they are generally in line with DOT CFR 192 and DOT CFR 195 guidelines. One of the supporting documents of CFR 192 says, "Through this required program, hazardous liquid operators will comprehensively evaluate the entire range of threats to each pipeline segment's integrity by analyzing all available information about the pipeline segment and consequences of a failure on a high consequence area."

This emphasizes the importance of following the four pillars of a good integrity program:

1. Identification of hazards
2. Collection of data
3. Detailed analysis
4. The consequence of failure.

The pipeline operator is mandated to set out a plan to collect all data required to perform the risk assessment. The plan must be able to prioritize the data collected for further analysis. Data should be collected for each threat identified for the pipeline system. In the beginning it must be assumed that all threats to the segment of pipeline are likely, and assessment should be made on that basis. There cannot be a standard list of data that can be collected, because the operation conditions of each pipeline are unique. Thus, the corresponding risk associated with them is also unique. However, to start, some perspective has to be established and data collected. The table below indicates the prescribed basic step to start collection of data for integrity management of a pipeline segment.

The source of data can be found in the construction data book that contains the process and instrument diagram (P&ID) drawings, as-built drawings, material data reports, inspection reports, hydro-test reports, weld maps, and reports on trenching, lowering and burial conditions, inspections reports, cathodic protection, and coating survey and inspection reports.

The existing management information systems (MIS) and geographic information system should be exploited to their full extent. Patrolling reports and aerial survey reports and photographs can be used to derive the conditions of specific pipe segments. Operating companies need to understand the importance of diligent data collection and record keeping. In my recent interactions with one of the leading gas pipeline companies in the USA, I have seen the important changes made by the great efforts of the Data and Record Integrity Supervisor and her tireless efforts to educate and train the field technicians in data collection review. The accuracy of information gathered, and information transmitted to drafting, is essential in preparing accurate as-built drawings.

Consulting with experts on the subject matter and conducting root cause analysis of the failures can also generate data that can be used for risk assessment and integrity management plans. The external sources such as databases from hydrology, demographic changes, population density, and variance should also be consulted.

The relevance of collected data must also be considered, for example, whether the data is related to the time dependent threats like corrosion and stress corrosion cracking (SCC). The collected data may not be relevant if it was collected several years ago. Stable and time independent threats are immune from time factors, however. The stable time dependent and time independent threats are discussed in subsequent chapters.

MAKING SENSE OF COLLECTED DATA

Data collection is not the end of the process; now that data are collected, there must be a system that can bring some meaning to the data collected from different sources.

The collected data must be brought together for analysis, and their context must be acknowledged for deriving correct interpretation of the data. Just like poor data input can present poor output of information, poor interpretation can also result in faulty decision making.

The collected data come from various sources and they are reported in various engineering units. To make them useful, a common reference system should be developed wherein all collected data are interpreted in the same common unit. This allows the data collected from different sources to be combined as one integrated whole. The station locations used to collect close interval survey data must be able to relate with other real collected data like in-line inspection (ILI), which is often reported in wheel counts. A method of using common engineering units should be developed to link the results of all important data types.

One example of such links might involve linking the report that indicates mechanical damage to the top of the pipe buried in a field, to a recent aerial photograph showing the farmer plowing the field. In another case, the periodic survey of cathodic potential indicates good CP coverage on a section of pipe, but corrosion on the pipe is suspected, so it is decided to conduct a coating condition survey by DCVG, which brings out the damaged coating location. It may be noted that different inspection and survey methods are not comprehensive one-tool-fits-all, but are complementary to each other. Their reports also need to be married to understand the full health status of a pipeline system. The knowledge of advantages and limitations of various inspection tools can be very useful here.

The initial structured evaluation of data referenced in Table 1-3-1 can point to the area of concern even without any inspection. Then, on the basis of this preliminary screening analysis, further and more detailed investigative inspections can be initiated and data collected for the segment of the pipeline. The use of pipeline aerial survey photography, GIS/MIS data, and potential impact area can start building a picture of the system. The tools used for assessing risk present their specialty reports that add another volume of data for analysis. Risk assessment tools are discussed in the next chapter.

Table 1-3-1 Prescribed Basic Elements of Data Collection.

Design data	Operational data	Inspection data	Construction data
Pipe wall thickness	Fluid quality	Pressure test	Year of installation
Diameter	Flow rate	In-line inspection (ILI)	Bending methods
Seam type and joint factor	MOAP	Geometry tool inspection report	Welding and inspection details
Manufacturer	Failure history	Bell hole inspection	Buried depth
Manufacturing date	Coating type and condition	CP inspection and data analysis	Crossing type (any cased crossings, etc.)
Material grade/ properties	CP system	Coating condition inspection (DCVG)	Pressure test
Equipment properties	Pipe wall temperature	Audit and reviews	Field coating type and application methods
	Pipeline inspection reports		Soil reports, back fill details
	Internal and external corrosion reports		Installation of CP system
	Pressure upheavals		Coating type
	Encroachments		
	Past repairs		
	Vandalism		
	External forces		

CHAPTER FOUR

Risk Assessment Tools

Contents

Pipeline Integrity Handbook
ISBN 978-0-12-387825-0

RISK ASSESSMENT TOOLS

When collected data is analyzed it will present such things as fault tree and event tree scenarios. These are the basic building blocks of risk assessment. These tools are used to capture the events and sequences of happenings just prior to the failure. They increase our understanding of the event and form the basis for a risk model.

FAULT TREE ANALYSIS

Developing a fault tree analysis (FTA) is a time consuming and costly process. The selective sub-system approach based on preliminary analysis is more useful. This allows dealing with much smaller systems; it also reduces the potential for errors. If required, the sub-system analysis can be integrated with other sub-system analysis to form the full system analysis.

The process starts with identification of the effect (failure) and the tree is then built from top to bottom, using each situation that could cause that effect. Failure probabilities are numbered with series of logic expressions. These numbers are the actual numbers about failure probabilities which are obtained from computer modeling programs that determine the failure probabilities.

The tree is usually written out using conventional logic gate symbols. The route through a tree between an event and an initiator in the tree is called a Cut Set. The shortest credible way through the tree from fault to initiating event is called a Minimal Cut Set.

Some industries use both fault trees and event trees, which is an inductive analytical diagram in which an event is analyzed using Boolean logic to examine a chronological series of subsequent events or consequences. An event tree displays sequence progression, sequence end states, and sequence-specific dependencies across time.

Five steps to FTA

Defining and understanding the undesired event

Definition of the undesired event can be very hard to catch, although some of the events are very easy and obvious to observe. An engineer with a wide knowledge of the design of the system or a system analyst with an engineering background is the best person to help define and number the undesired events. Undesired events are then used to make the FTA, where one event per FTA is developed.

Understanding of the system

Once the undesired event is selected, all causes with probabilities of affecting the undesired event of zero or more are studied and analyzed. Getting exact numbers for the probabilities leading to the event is usually difficult. Computer software is used to study probabilities; this may lead to less costly system analysis.

System analysts can help with understanding the overall system. System designers have full knowledge of the system and this knowledge is very important for not missing any cause affecting the undesired event. For the selected event, all causes are then numbered and sequenced in the order of occurrence and are then used for the next step, which is drawing or constructing the fault tree.

Construction of the fault tree

After selecting the undesired event and having analyzed the system so that all the causing effects and their probabilities are known, a fault tree can be constructed. A fault tree, as stated above, is based on "AND & OR" gates, which define the major characteristics of the fault tree.

Evaluation of the constructed fault tree

After the fault tree has been constructed for a specific undesired event, it is evaluated and analyzed for any possible improvement. At this step all possible hazards that may affect the system are identified. This step is an introduction to the final step, which will be to control the identified hazards.

Control the hazards identified

This is a very system-specific step, as for every individual system there will be a different set of hazards to identify. However, the key point is that, as the hazards are identified, all possible methods are pursued to decrease the probability of their occurrence.

BASIC MATHEMATICAL FOUNDATION

As is evident from the above description, the FTA and event identification are mathematical tools for identifying the hazards and risk assessment. The events in a fault tree are associated with statistical probabilities. For example, component failures typically occur at some constant failure rate λ (a constant hazard function). In this simplest case, failure probability (P) depends on the rate λ and the exposure time t:

$$P = 1 - \exp(-\lambda t)$$

$$P \approx \lambda t, \lambda t < 0.1$$

A fault tree is often normalized to a given time interval. Event probabilities depend on the relationship of the event hazard function to this interval.

The development of such a Boolean logic-based program is done by computer and mathematical model developing experts. Integration of such experts in the risk assessment team will ensure proper FTA.

Comparison of FTA with FMEA

FTA is a deductive, top-down method aimed at analyzing the effects of initiating faults and events on a complex system. This contrasts with failure mode and effect analysis (FMEA), which is an inductive, bottom-up analysis method aimed at analyzing the effects of single component or function failures on equipment or sub-systems. FTA is very good at showing how resistant a system is to single or multiple initiating faults. It is not good at finding all possible initiating faults.

FMEA is good at exhaustively cataloging initiating faults and identifying their local effects. It is not good at examining multiple failures or their effects at a system level. While FTA considers external events, FMEA does not. In a good risk assessment program it is good practice to adopt both tools and use the failure mode effects summary to interface the two systems.

CAUSE AND EFFECT (ISHIKAWA) DIAGRAMS

Cause and effect (Ishikawa) diagrams (Figure 1-4-1) are causal diagrams that show the potential factors causing a certain event. Each possible cause or reason for imperfection is a

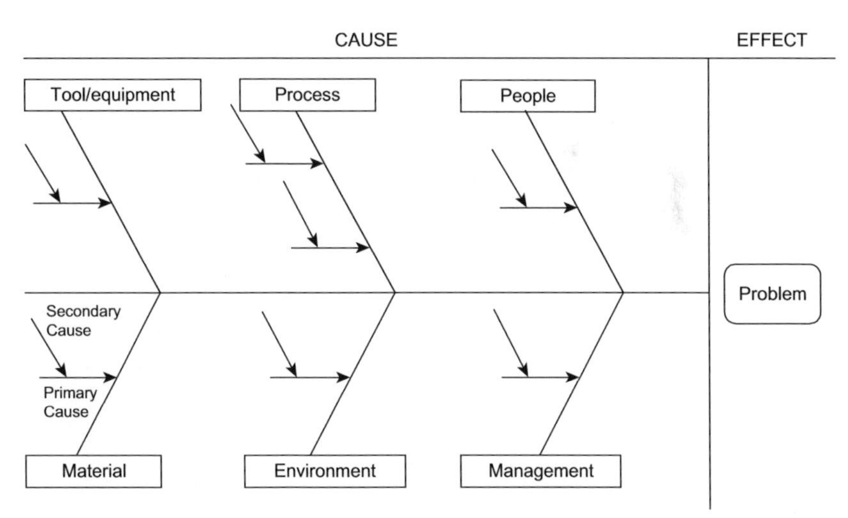

Figure 1-4-1 Ishikawa cause and effect diagram.

source of variation. Causes are usually grouped into major categories to identify these sources of variation. The categories typically include:

1. People: education, training, and experience of the people performing the task
2. Methods: how the process is performed and the specific requirements for doing it, such as company policies, procedures, rules, and regulatory requirements
3. Machines: tools and equipment required to accomplish the job
4. Materials: selection of material and parts used to construct and run the pipeline system
5. Measurements: data generated from periodic maintenance and inspections
6. Environment: the conditions, such as location, time, pressure, temperature, and corrosivity of the fluid.

Causes

Causes in the diagram are often categorized according to the "8 Ms" described below. Cause and effect diagrams can reveal key relationships among variables, and the possible causes provide additional insight into process behavior.

Causes are often derived from idea generation sessions. These groups can then be labeled as categories of the fishbone:

1. Man (operators, qualifications, and training)
2. Materials (grade, quality of fabrication, and installation)
3. Methods (processes)

4. Machines (technology)
5. Measurements (inspection)
6. Mother nature (environment and natural causes)
7. Management (commitment, resources)
8. Maintenance (type, frequency, philosophy).

These typical 8Ms are often associated with the production process. They will typically be one of the traditional categories mentioned above, but may be something unique to the application in a specific case. Causes can be traced back to root causes with a questioning process. A typical tool that is used for this purpose is often referred as the "5 Whys," the objective of which is to get to the root of the cause and not stop at the first available symptom.

While developing a cause and effect diagram, the typical questions that may be asked for each group are given below. These questions should aim to extract information and may be suitably modified to accommodate a specific task in mind.

People

Was the scope of the work properly interpreted?
Were the correct drawings and work-related specifications issued for the scope of work?
Did the recipient understand the information?
Were people who were assigned to the task given proper training aimed at the performance of the task?
Were instructions easy to understand or was too much assumed to be understood?
Were the decision-making guidelines available?

Did the environment influence the actions of the individual?
Were there distractions in the workplace?
Was fatigue a mitigating factor?
How much experience did the individual have in performing this task?

Machines

Was the correct tool used?
Was the equipment affected by the environment?
Was the tooling/fixturing adequate for the job?
Was the equipment being properly maintained (i.e., daily/weekly/monthly preventative maintenance schedule)?
Did the machine have an adequate guard?
Was the equipment used appropriately within its capabilities and limitations?
Were all controls, including emergency stop buttons, clearly labeled and/or color-coded or size differentiated?
Was the equipment the right application for the given job?

Where data is collected and filed, the following additional questions may be added to the list:

Did the equipment or software have the features to support the needs/usage of the project?
Was the machine properly programmed?
Were files saved with the correct extension to the correct location?
Did the software or hardware need to be updated?

Measurement (inspection and testing)

Did the gauge have a valid calibration date?

Was the proper gauge used to measure the part, process, chemical, compound, etc.?

Was a gauge capability study ever performed?

Did measurements vary significantly from operator to operator?

Did operators have a tough time using the prescribed gauge?

Was the gauge fixturing adequate?

Did the gauge have proper measurement resolution?

Did the environment influence the measurements taken?

Materials (including raw material, consumables, and information)

Was all needed information available and accurate?

Can information be verified or cross-checked?

Has any information changed recently? Do we have a way of keeping the information up-to-date?

What happens if we do not have all of the information we need?

Is a Material Safety Data Sheet (MSDS) readily available?

Was the material properly tested? What constitutes a proper test?

Was the material substituted? What was the basis for the substitution?

Was the supplier's process defined and controlled?

Were quality requirements adequate for part function?

Was the material contaminated?

Was the material handled properly (storage, dispensing, usage, and disposal)?

Environment

Was the process affected by temperature changes over the course of a day?

Was the process affected by humidity, vibration, noise or lighting?

Did the process run in a controlled environment?

Were associates distracted by noise, uncomfortable temperatures, fluorescent lighting, etc.?

Methods (processes)

Was the material properly identified?

Were the workers properly trained in the procedure?

Was the testing performed statistically?

Was data tested for true root cause?

How many "if necessary" and "approximately" phrases are found in this process?

Has a capability study ever been performed for this process?

Is the process under Statistical Process Control (SPC)?

Are the work instructions clearly written?

Are mistake-proofing devices/techniques employed?

Are the work instructions complete?

Is the tooling adequately designed and controlled?

Is handling and storage adequately specified?

Was the process changed?

Was the design changed?

Was a process FMEA ever performed?

Was adequate auditing conducted?

Are hazardous operation analyses conducted and hazards identified and publicized to personnel?

Ishikawa diagrams identify the necessary conditions that may lead to the failure, but it falls short of making the distinction between necessary conditions and sufficient conditions. There should be sufficient conditions to lead to actual failure.

HAZARD AND OPERABILITY (HAZOP) STUDY

Most pipeline operators conduct HAZOPs of pipeline along with design prior to the start of construction. The information available through study is an important source of data for risk assessment. HAZOP is a multi-disciplinary activity conducted through sets of meetings, with these meetings progressing through the HAZOP guidewords (Table 1-4-1) from the multi-disciplinary team members. The approach of the technique is qualitative.

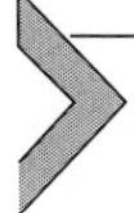

FAILURE MODE AND EFFECT ANALYSIS (FMEA)

Somewhat similar to HAZOP is FMEA. This is a tool that identifies potential failure mode within a system. This two-part study system first identifies the errors and defects of a system (the failure mode) and the second part (the effect analysis) identifies the consequences of such failure. The tool classifies the failure by severity and likelihood.

Table 1-4-1 HAZOP Guidewords.

Parameter / guide word	More	Less	None	Reverse	As well as	Part of	Other than
Flow	high flow	low flow	no flow	reverse flow	deviating concentration	contamination	deviating material
Pressure	high pressure	low pressure	vacuum		delta-p		explosion
Temperature	high temperature	low temperature					
Level	high level	low level	no level		different level		
Time	too long/too late	too short/ too soon	sequence step skipped	backwards	missing actions	extra actions	wrong time
Agitation	fast mixing	slow mixing	no mixing				
Reaction	fast reaction/ runaway	slow reaction	no reaction				unwanted reaction
Start-up / Shut-down	too fast	too slow			actions missed		wrong recipe
Draining / Venting	too long	too short	none		deviating pressure	wrong timing	
Inserting	high pressure	low pressure	none			contamination	wrong material
Utility failure (instrument air, power)			failure				
DCS failure			failure				
Maintenance			none				
Vibrations	too low	too high	none				wrong frequency

The table gives an overview of HAZOP guidewords, parameters, and their interpretations.

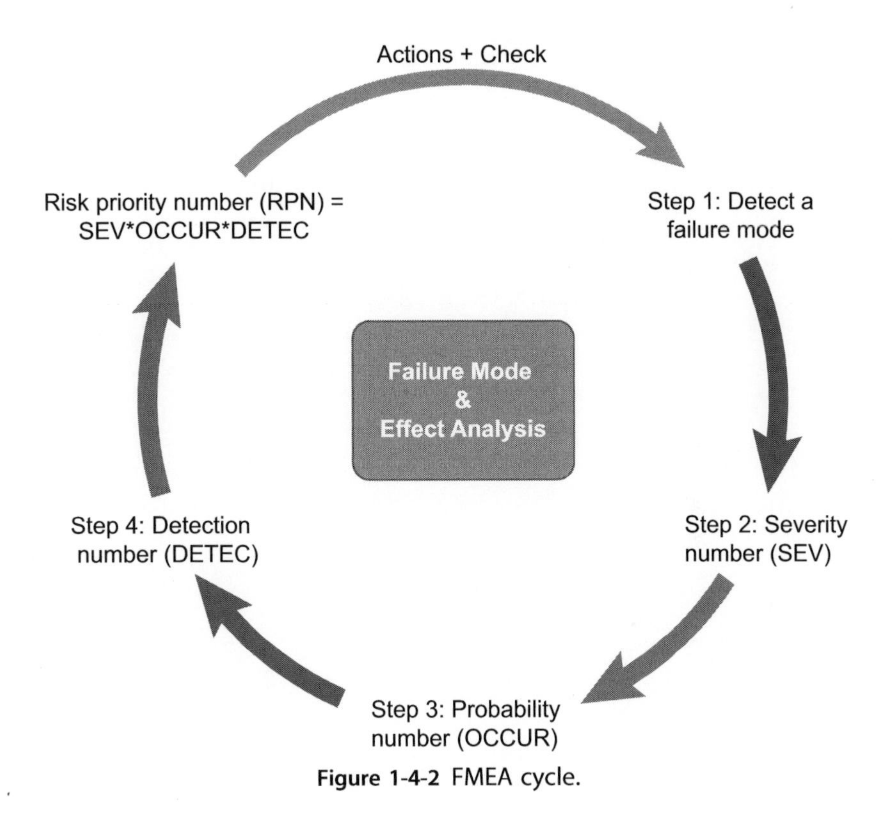

Figure 1-4-2 FMEA cycle.

FMEAs are very useful tools, especially when you extend your risk assessments to surface facilities like tank farms and pump stations (Figure 1-4-2). As stated above, these are tools (components) of a complete risk model.

For a proper study of risk, the system is often divided into segments. These segments are studied individually. The segmenting is often done on the basis of change in risk variable. Another alternative is to use fixed interval segmenting, for example, segmenting at each MLV site, each compressor station, or each mile.

The advantage of using variable risk segments is that each segment represents a possible risk and equal attention is given to each risk potential.

CHAPTER FIVE

Hazards and Threats to a Pipeline System

Contents

Pipeline Integrity Handbook
ISBN 978-0-12-387825-0

THREATS LEADING TO RISK

Definition of risk

- **The probability of an event that causes a loss and the potential magnitude of that loss.**
- **Example: Failure in oil & gas pipelines can release causing damage. In an unpopulated area the damage is slight but in dense population centers the consequence is high; these are termed as high consequence areas (HCAs).**

From the preceding chapter we can derive that it is important to understand what the threats are that raise the risk probability in a pipeline system and how to assess them. Generally, there are about nine major identified threats to a pipeline system, which can be categorized into three groups. The first group of hazards are time dependent. This means that they occur after some time in service. The second group of hazards are stable, signifying that they are present irrespective of time in service. The third group is classified as time independent. These all are briefly discussed in this chapter.

External corrosion threat

Definition of hazard

- **A characteristic or group of characteristics that provides the potential for a loss.**
- **Examples: Internal and external corrosion, stress corrosion crack, material defects, third-party damage and environment including toxicity, flammability, exposed pipes, lack of signage, etc.**

Pipeline Corrosion Control

- **Coating is the main corrosion control technique.**
- **But dependence on one source for protection is not a very wise approach - No coating is perfect.**
- **To protect coating flaws, cathodic protection (CP) is applied.**
- **Coatings and CP are successful combination for corrosion control.**

External corrosion on a pipeline could be microbiologically induced corrosion (MIC), or from galvanic action or electrochemical reaction, where the steel pipe becomes an anode in an electrochemical cell. This is a time-dependent hazard, and occurs after some time in service.

To address these threats, collection of data is carried out, and based on the analysis of those data, a risk assessment plan is prepared. This is an important step. The data collection process is a painstaking and detailed activity. In the beginning, the collected data may be overwhelming and intimidating, but a good risk assessment program is data dependent. In fact, most of these data are already in the computers of the pipeline operators but they keep sitting as dead data. In the world of computers, it is easy to share, download, and transfer most of these valuable data and put them to better use. "Better use" is the availability of the data for analysis that helps manage the risk. Periodically this collection of data can be refreshed and updated. There is no limit to the available data and its possible use. However, the basic data that needs to be collected in this context is listed below:

a. Pipeline year of installation
b. Pipe grade (ASTM A 106 Grade B, API 5L Grade B, X42, X65, X70, etc.)
c. Pipe diameter and wall thickness
d. Maximum operating pressure (MAOP) (or design pressure) and temperature
e. Pipe seam type, e.g., electric resistance welding (ERW), seamless, double submerged arc welding (DSAW)
f. Coating type, e.g., coaltar enamel, fusion bonded epoxy (FBE), three-layer polyethylene (3LPE)
g. Coating condition at the time of last inspection (updated periodically)

h. Cathodic protection installed, its history of data collection and current status
i. Soil characteristics (type, resistivity, physical attributes)
j. History of inspection, inspection methods, reports, and concerns
k. MIC detected (any history)
l. Leak history (time, extent, cause, corrective action taken)
m. Past hydrostatic test information

For new pipeline systems most of this data is readily available, as the details of original material selection and design criteria are collected as baseline data. The as-built drawing with material information assumes critical importance. The input to as-built drawing should be carefully reviewed by a responsible engineer before finalizing it as ready for the draftsmen to enter the data on the drawing. A well-built drawing is likely to prevent problems and guess work in the subsequent life of the pipeline system. A well-established data and record integrity team within the operator's engineering division should lead such data collection and review efforts. The success of this is briefly discussed in Chapter 3, as it has enormous bearing on the integrity of the pipeline system.

The probability of external corrosion threat can be assessed by the use of inspection tools such as inline inspection (ILI) by using a smart pigging system; this can be supplemented with inspection methods such as pressure testing and external corrosion direct assessment (ECDA) tools.

ECDA

Pipeline External Corrosion Control -1

Pipeline coatings form a barrier between the corrosive moisture in the soil surrounding the pipeline and the metal pipe. There are many types of coating applied both in the factory and in the field. Field welds are coated by field joint coating systems.

External Corrosion-2 : Cathodic Protection

Pipeline corrosion occurs when current flows off of a pipeline. Using impressed current, a CP system induces current into the ground by using a rectifier and ground bed to stop the flow of current off of the pipeline and protecting areas of the pipe not protected by coating.

ECDA is a structured process that is intended to improve safety by assessing and reducing the impact of external corrosion on a pipeline. Other tools for assessing external corrosion are direct current voltage gradient (DCVG), close interval potential (CIP), ILI using intelligent pigging, and pressure testing.

Use of guided wave ultrasonic testing (GWUT) as an alternative tool for ECDA has been promoted and sometimes used. A study was conducted by the Gas Technology Institute (GTI), Pipeline Hazardous Material Safety Administration, Department of Transportation (PHMSA/DOT) and a report published in August 2008. A brief introduction to the objectives and methodology of the study is included here in this book, and people interested in reading further details of the study are encouraged to obtain the full report.

GWUT testing method

The Gas Technology Institute in collaboration with DOT (Project number 195) conducted a study to evaluate the applicability and reliability of GWUT testing methods. The project stakeholder group reviewed the ECDA demanding situations from *2005 PHMSA R&D Forum* and previous research activities. They agreed to and volunteered the following three high priority situations to focus on for potential case studies.

Multiple Pipes (Structures) in a Congested Right of Way

Interference issues with above ground inspections; stray currents; complex meter and station piping.

Bare Pipe Segments

Cased Crossings Industry needs better differentiation between metal loss and casing/pipe contact points. Sizing of defects inside casings; uncased crossing and deep crossing situations; long crossings (e.g., use pitch-catch vs. pulse-echo GWUT).

The following tools were used during the integrity assessments performed during this project:

1. GWUT (GUL and Teletest): torsional and longitudinal signals, pitch-catch and pulse-echo, C-can, and multiple frequency ranges
2. Magnetic tomography inspection
3. Visual inspection
4. Manual and Porta-Scan UT
5. Radiography (X-ray)
6. Magnetic particle inspection (MPI)
7. Close interval surveys (CIS)
8. Direct current voltage gradient (DCVG)
9. Pipeline current mapper (PCM), native potential and side-drain surveys
10. Soil resistivity

These three situations with ten different tools resulted in 30 excavations for GWUT application and when combined with the in-kind data, included a total of approximately 100 dig sites with 55 confirmed (a 100% validation) indications for analysis.

All validated data was collected, analyzed, and summarized in graphical form, which included the following steps:

- Inspection ranges
- Confirmed defect sizes (depth, length, width, and volume)
- Probabilities of detection (both false/true positives and negatives)

Continued

Some general lessons learned in each of the three cases are listed below.

For Multiple Pipes (Structures) in Congested ROW Situations

A. ECDA standard tools worked well in open areas where interferences did not preclude the use of CIS, DCVG, and PCM as validated by 100% excavation with visual inspection and pit gauge and MPI.
B. GWUT was very effective when standard DA tools could not be used. GWUT also identified the presence of sludge and deposits in pipe sections.

For the Bare Pipe Situations

A. CIS coupled with native potential surveys and side-drain surveys (also known as hot spot surveys) worked well and predicted areas of potential past corrosion.
B. GWUT had a relatively short range due to the very adherent and "plastic" clay soil.
C. Magnetic tomography did not correlate well (false positive indications) for corrosion but did locate a wrinkle bend type feature outside of the GWUT inspected section.

For Cased Pipe Situations

A. GWUT correlated with the direct exam findings.
B. For thick, pliable, well-adhered asphalt coatings, the GWUT range was severely restricted.

C. PCM inspections provided another means of determining short situations between carrier and casing pipes.

All of these lessons, and many more learned from this project, were compiled and are presented as a report, Guided Wave Ultrasonic Testing Background, Technical Explanation, and Field Implementation Protocol to Assist Operators. *For full details of this study, it is recommended that readers obtain and read this report.*

On the same subject, a paper entitled *Feasibility Study Relating Guided Wave Ultrasonic Testing (GWUT) and Pressure Testing for High Pressure Steel Pipelines*, proceedings of IPC 2008 and paper number 64403, was presented in Calgary, Alberta, Canada, at the 7th International Pipeline Conference, September 2008, by K. Leewis-P-PIC, D. Erosy of GTI and G. Matocha of Spectra Energy. The paper concluded the following:

- Both amplitude and directionality are required to estimate a defect magnitude.
- Major equipment manufacturers already provide both the amplitude and directionality in their reporting information.
- The feasibility of GWUT equivalence to a hydro-test has been shown, with the understanding that further validation is needed.

These other tools for external corrosion assessment are discussed in this chapter.

ECDA is a continuous improvement process, whose application allows a pipeline operator to identify and address the location where external corrosion is actively occurring or may occur. One of the key advantages of ECDA is that it can identify the location where external corrosion may occur in the future. Comparison of successive ECDA reports can also demonstrate changes in the pipeline integrity status.

Coatings – Field Joint
Coating of Pipeline Weld

The ECDA process involves collection of data to determine whether ECDA is feasible, to define ECDA regions, and select indirect inspection tools. This step in the ECDA process is often termed as a pre-assessment step.

The pre-assessment is followed by an indirect inspection. At this stage, the inspection is carried out from ground surface to determine the severity of coating faults and other anomalies. Often two or more inspection tools are applied to cover the pipeline segment. The next step is the direct examination.

Direct examination includes analysis of data collected from indirect inspection. This may lead to identification of locations for excavations (bell holes) and pipe surface examination. At this stage other aspects of pipeline, such as coating and surface corrosion, are also evaluated. The evaluation leads to the mitigation of these anomalies.

The fourth step involves the evaluation of data from three earlier steps and assesses the effectiveness of the ECDA process.

Coatings – Epoxy Coated Weld Joint

Weld Joint Before Coating

Coating Integrity

Pipeline Coating Failure

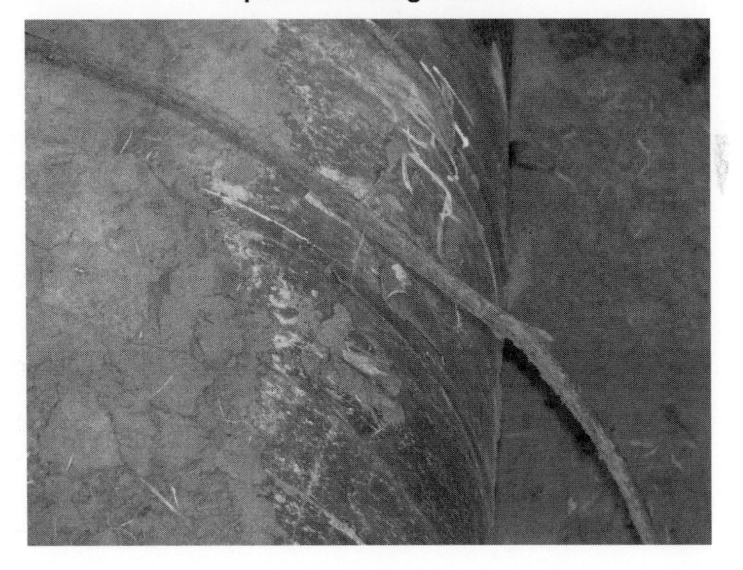

Pipeline Coating Damages

Pipeline Coating Damage

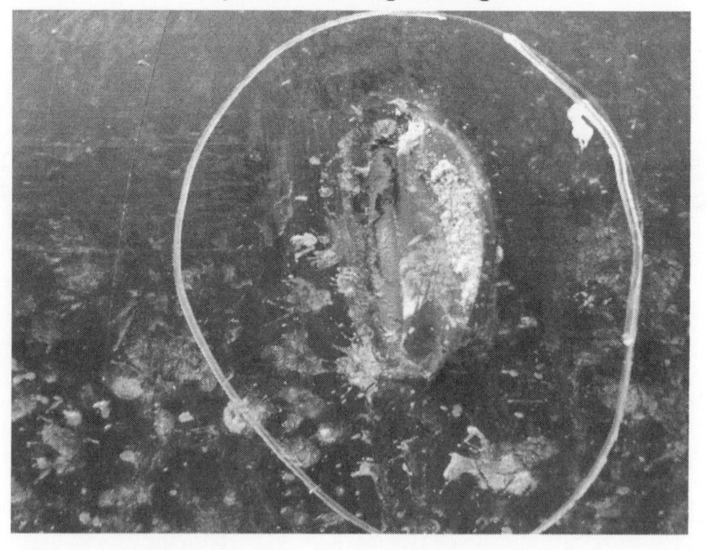

Pipeline Coating Damage

Assessment of Coating Failure

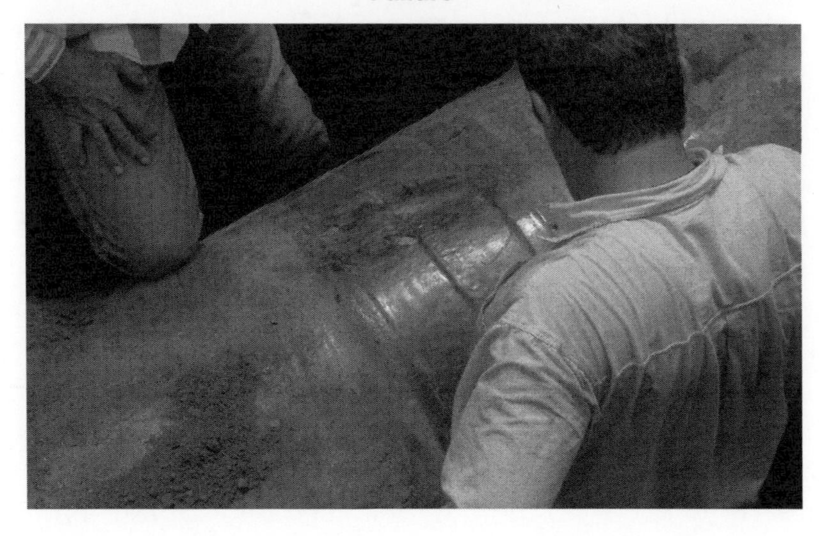

Coating Inspection

Two other tools used to determine the condition of the pipeline and improve its safety are CIP survey and measurement of the DCVG along the pipeline segment. These are discussed below.

CIP survey

The CIP survey provides a detailed profile of the potential level along the entire route of a pipeline. This profile can be used to assess the performance of the CP system and also provides information on the coating system and interaction effects. The route is identified and pre-marked, or tools are used to locate the pipeline route immediately ahead of the survey operator. The location is often correlated to GPS positioning markers to establish proper identification.

CP CIP Survey

The following basic tools are required for conducting a survey:

a. Pipeline location equipment.
b. $Cu/CuSO_4$ reference electrodes fitted at the tip of poles.
c. Microprocessor-controlled data collector or voltmeter with a minimum input impedance of 10 Mohms and sufficient level of AC rejection.
d. Synchronized timers or current interrupters for switching DC power sources.
e. Insulated wire, which may be disposable or re-usable, wound on a dispenser.
f. Distance recorder.

To obtain polarized potentials, synchronized current interrupters are installed in all transformer rectifiers or other DC power sources providing CP current to the pipeline.

As stated earlier, where there is potential to face interference or fluctuating potentials, stationary recording units are installed.

An insulated wire is connected to the pipeline CP test post. The survey proceeds with the operator carrying the mobile data collector and walking with the poles with $Cu/CuSO_4$ reference electrode touching the ground above the pipeline. The positioning of electrodes and reading is collected at an interval of 1 to 1.5 meters. For reference purposes, the features along the route are recorded into the data recorder and they help in interpretation. Survey sections between connection points to collect data and analyze it are kept short.

CIP surveys are carried out to identify the levels of protection that exist along pipelines. It is common practice among pipeline operators to use CIP as compared to the measurement of potentials at the test points. This is a more comprehensive tool in assessing the condition of the entire length of pipeline. The process includes trailing wire that is connected to the closest test point and potentials are taken at intervals of typically 1.5 meters. The measurement includes an IR voltage drop in the reading caused by the flow of CP current in the soil. To overcome this error, timers are fitted in the DS circuit of the transformer rectifiers or other power supplies to interrupt the protection current. Where protection is provided by two or more sources, the timers are synchronized to interrupt all current sources simultaneously. Potentials are recorded immediately after interruption but before significant depolarization occurs. Readings taken in this way result in IR voltage being drop free. This is called polarized potential and is also known as instantaneous-OFF potentials. A CIP survey gives the following detailed information about the pipeline:

- The entire pipeline section is walked. This helps to inspect all the CP equipment and the right of way at the same time as the survey.
- Provides a complete pipe to soil potential profile on the pipeline for both ON and polarized-OFF potentials.
- Identifies areas of stray current interaction; where pipe to soil potential fluctuates, stationary data collectors are used at chosen test locations.
- A hard copy of the survey data is produced allowing easy identification of defect areas by non-technical personnel.

- A baseline survey of the pipeline potentials can be obtained, providing guidance for future operation and maintenance of the CP system.
- Provides information on CP levels on coating defects and likely active corrosion location.
- Can easily be combined with integrated GPS and DCVG measurements.

Most CIP systems on the market have the facility to let the raw data be transferred to computers; at the end of each daily survey the data is downloaded, checked, and analyzed. Proprietary programs are now available that can analyze the data and present them in graphs and charts.

DCVG survey

As in CP, if a DC current is induced on a pipeline, the ground voltage gradients are created due to the passage of the current through resistive soil. If pipe coating is good then it will have good resistance to these currents. However, if the coating has breakage, called a coating "holiday," then at these locations the resistance will be low and the current will flow through the soil and be picked up in the pipe at holiday locations. A measurable voltage gradient will be created at the adjacent ground. The larger the defect, the greater would be the current flow. Increasing the current flow also results in an increased voltage gradient in a given soil resistivity. Survey techniques can provide an assessment of coating condition over areas that are difficult to access, i.e., road, rail, and water crossings.

With the help of special interrupters, pulsed frequency DC current is applied to a pipeline. This distinguishes the response of a coating defect from that of stray traction and telluric currents. Often the existing CP system is used to introduce the required signal; if no CP system is available to the pipeline then temporary earths are established at convenient connection points along the line.

The defects found during a DCVG survey are sized and mapped by GPS coordinates and are fully documented on special graphical reports.

Coating – DCVG Survey

DCVG surveys may be carried out to determine any or all of the following:

1. Locating coating defects
2. Sizing of coating defects

3. Progressive monitoring of coating
4. Assessing the available level of CP and coating protection
5. Prioritizing the repair and maintenance activity
6. CP system adjustment or upgrade

DCVG survey process

Prior to the commencement of any survey section, a current interrupter is installed in the nearest existing CP station or temporary current source which may be established as necessary. Typically, a minimum potential swing of 500–600 mV is desired, so the current source output is adjusted to meet the potential shift requirements. The following is the step-by-step survey process:

1. At the start, the difference between "on" and "off" potentials is recorded at the test point nearest the survey start point. This step is repeated at all other test points encountered on the way and the survey is commenced.
2. The operator traverses the pipeline route using the probes as walking sticks. One probe is in contact with the ground at all times and for a short duration between strides both probes must be in ground contact. One probe is kept on the centerline of the pipeline and the other is maintained at a lateral separation of 1–2 m or probes can leapfrog along the centerline. ***If no defects are present, the needle on the voltmeter registers no movement.***
3. As a defect is approached a noticeable fluctuation is observed on the voltmeter at a rate similar to the interruption cycle. The amplitude of the fluctuation increases as

the defect is approached and voltmeter sensitivity is adjusted as necessary. If the probes are maintained in a similar orientation parallel to the pipeline, the swing on the voltmeter is directional.

4. This allows the defect to be centered by detailed maneuver around the epicenter. Using the data obtained by DCVG, each defect can be analyzed. They can be sized by relating the potential swing, read as voltage, to remote earth (mV1) to the signal voltage recorded at the nearest two test posts mV2 and mV3. The distances of defect to these two test posts (m1, m2) are also brought into account. In addition, it is also possible to determine whether active corrosion is taking place at the defect.

Data analysis

Coating defects are recorded on reports with reference to a fixed reference point marked on the route alignment sheets and a stake placed in the ground.

Comments on signal strength are also recorded and the defect graded as follows, where:

$$\%IR = mV1/\ \{mV2 - (m1/(m1+m2) * (mV2 - mV3))\}$$

1. Values greater than 35% IR indicate large and serious defects that require immediate attention.
2. Values between 16% and 35% IR indicate medium defects that require attention and can be addressed during general maintenance.
3. Values less than 15% IR are minor defects and need not be repaired but their behavior should be further monitored.

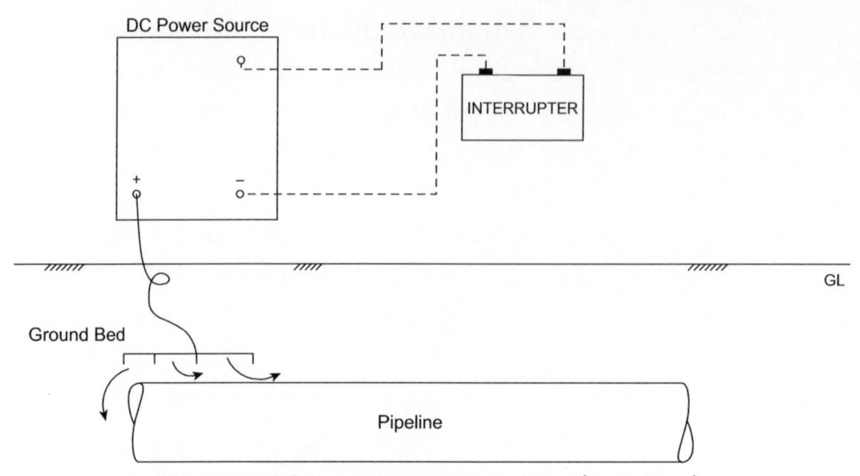

Figure 1-5-1 Interrupter set-up to introduce signal.

Internal corrosion threat

Internal corrosion may be either due to chemical reaction to the pipe internal surface or material loss due to micro-biological reactions, which is also an electrochemical reaction Figure 1-5-1.

For internal corrosion assessment, apart from common points such as pipe material diameter, year of installation, leak history, additional points to be considered and data collected on include the following:

- Fluid chemistry, including the presence of free water, oxygen, hydrogen, sulfur, hydrogen sulfide, chlorides, carbon dioxide, etc.
- Assessing the reaction of the gases, liquids, and solids with pipe material in the given temperature and pressure conditions, and flow velocity.
- Bacterial culture and operating stress level.

It is important to note that although the external coating and CP system is not directly involved in protection from internal corrosion, its role in the overall integrity of the pipe system is significant and cannot be ignored.

Measurement and monitoring of internal corrosion

Measurement of internal corrosion, and the methods used to obtain and gather these measurements, have often left the operators of internal corrosion monitoring systems frustrated because of the complicated and often difficult procedures involved. The current intrusive monitoring methods and the first generation of patch or external type devices have increased the cost of internal corrosion monitoring programs.

The latest developments in equipment, instrumentation, and software for control and measurement of internal corrosion offer the operators an accurate and user-friendly alternative to what had been previously used. The instrumentation and software, by bringing in other parameters that cause the onset or increase in corrosion, now allow the operators to relate corrosion to the events that cause it.

Basic internal corrosion monitoring techniques

The basic internal corrosion monitoring techniques include the following:

1. Coupons (C)
2. Electrical resistance (ER)
3. Linear polarization resistance (LPR)
4. Galvanic (ZRA)
5. Hydrogen probes and patches (H2)

In addition to the above, there are some new corrosion monitoring techniques that have been tested in the laboratory and in the field:

1. Surface activation and gamma radiometry
2. Impedance (EIS)
3. Electrochemical noise (ECN)
4. Acoustic emission
5. Real-time radiography
6. Real-time ultrasonic
7. Hydrogen patches (H2)

A paper on the subject has been presented by Gerald (Jerry) Brown of Brown Corrosion Services, Inc., Houston, TX, USA. This paper is reproduced below with permission for more detail on the subject. The paper discusses various types of internal corrosion and failure. The paper also describes different types of corrosion monitoring technology and some recent examples of field data.

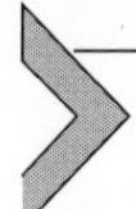

CAUSES AND RATES OF INTERNAL CORROSION

The causes of internal corrosion are many and can be generally divided into the following areas Figure 1-5-2:

1. The chemical composition of the stream
2. The physical factors of the stream
3. The physical factors of the structure

Chemical composition of the stream

Factors that affect the corrosion are:

- H_2O content
- H_2S content
- CO_2 content
- Dissolved solids
- Organic and inorganic acids
- Elemental sulfur and sulfur compounds
- Bacteria and its by-products
- Hydrocarbons
- pH
- Interactions of all of the above, other trace elements, and chemistry variables

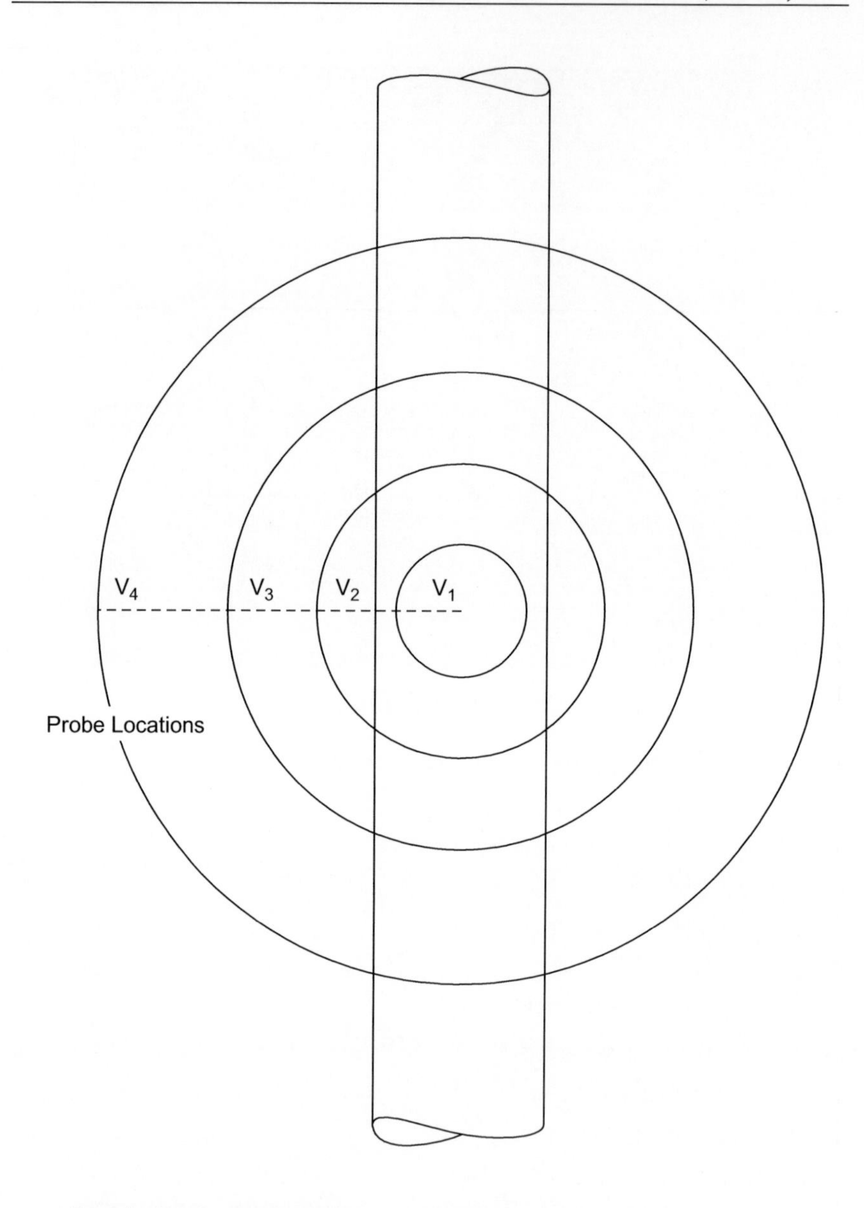

Over the line to remote earth
Voltage drop = $V_1 + V_2 + V_3$ ...... V_n reading

Taking OL/RE voltage drop
Figure 1-5-2 Reading OL to RE voltage drop.

Physical factors of the stream

Factors that affect the corrosion are:

- Temperature
- Pressure
- Velocity
- Vibration
- Entrained solids and liquids
- Deposits
- Flow characteristics and patterns (slugs)
- Interaction of all of the above and other physical factors of the stream

Physical factors of the structure

Factors that affect the corrosion are:

- Materials of construction
- Residual/operating stresses
- Design factors
- Crevices
- Depositions
- Surfaces

Most importantly, however, the interactions of all of these above factors can cause corrosion. As you can see with the interaction of all of these factors, plus others, the ability to predict the corrosion rate or the hydrogen permeation rate becomes, at best, very difficult.

Dynamic systems are always changing and, therefore, corrosion types, rates, and locations change. The question is often asked why do the systems change, and why do we not stabilize the system variables, thus controlling the corrosion and, therefore, also controlling the internally related corrosion failures. The simple answers go back to the inlet products that are used and the fact that different feedstocks must be used which all have different characteristics. In addition, the flow from the original oil and gas wells changes over time as the water cut, chemistry, pressure, and temperature changes occur during the depletion of the reservoirs that hold these fluids and gases. Of equal importance is the fact that to operate cost effectively today different products must be placed in the pipelines to economically justify the use of existing facilities and the building of new ones.

In pipeline operations the product also changes over time. In addition, the economics of running pipelines now calls for transporting several different products through the pipelines. By mixing products, new situations arise where corrosion rates can be greatly increased. We are also faced with the situation where pipelines may be in the ground so long that slow corrosion rates can cause failures that we have not previously seen.

For all practical purposes, there will be no corrosion without water. Water, if only a thin film, is found in most petroleum-based systems throughout the world. In addition the sporadic wetting and drying that takes place during slug flow often exacerbates the corrosion mechanism. Water collection at certain parts of piping systems also leads to isolated corrosion cells.

METHODS OF CORROSION CONTROL

All methods of corrosion control fall into one of the following categories:

1. Coatings
2. Cathodic protection (anodic protection)
3. Filming inhibitors
4. Alteration of the environment
5. Material selection
6. Alteration of the structure
7. Repair or replace

However, whatever method, or combination of methods, of corrosion control one uses, corrosion monitoring must be in place. As an example, if you are using coatings for corrosion protection, when do you know if the coatings have failed or how do you determine if the coatings are still protecting the surface? If the environment is altered by raising the temperature, how does one know if he has approached the onset or threshold of corrosion? If a pipeline dead-leg is in the system, how does one know if the corrosion in the dead-leg is non-existent or accelerated? Internal monitoring can give you all of the answers.

Monitoring versus inspection

Inspection is a very important and necessary tool. However, any inspection program or inspection project, regardless of the method used, will only give you data on the metal deterioration between two or more specific points in time when the inspection was performed. Also of most importance is the fact that all the data that is gathered is from the past. This means that if corrosion control measures are implemented, for example, increased doses of inhibitor or removal of water, the basis of choosing these actions is based upon the parameters as they were in the past, not as they are when the corrosion control measures are implemented. Also, the results of these corrosion control measures will not be observable until the next time of inspection.

By basing your corrosion control measures upon inspection rather than monitoring, you may not be able to actually stop or lower the corrosion rates, which are the most common reason to monitor in the first place.

Monitoring, on the other hand, on a near real-time basis, will give you data on what is happening now.

By getting near real-time data you are able to relate corrosion data to events that caused the corrosion; therefore, you will be able to lower corrosion rates by adjusting or eliminating the events that caused the corrosion to occur. Examples of obvious events causing corrosion could be the shutdown of a chemical injection pump, wash water escaping into the system, or a leaking pump sucking air (oxygen) into a closed liquid system. Once the upset condition is correlated with the increase in corrosion rate, these events can be stopped or modified, so that they do not cause corrosion.

Inspection, however, should not be overlooked, but used in conjunction with monitoring. One of the most advantageous factors when using some inspection techniques such as intelligent pigging is that this type of inspection can cover virtually whole piping systems. Monitoring does not cover the whole system and even the best engineering can miss the actual spots where corrosion may be occurring.

When the inspectors complete a project and leave the site, the question remains: who is monitoring the system? A well-planned monitoring program should be your sentinel until the next inspection.

Corrosion monitoring techniques

Several basic corrosion monitoring techniques have been used for many years. All of these methods have their place and work well in the proper locations and applications. Most of these traditional methods require intrusion into the stream to be monitored. This intrusion is necessitated because the sensor must be exposed to the environment to be studied. By making this intrusion, four general considerations should be of concern:

1. A hole of some sort must be made in the pipe and/or vessel.
2. The sensor is not of the same material as the structure to be monitored, nor is it in the exact same location where corrosion is occurring.
3. A probe changes flow patterns that, in turn, can determine different corrosion rates.
4. The surface areas of the sensors are restricted to very small areas.

The traditional monitoring methods have undergone drastic changes over the last several years due to the ability the computer has given us to digest large amounts of data and display this data in such a way that it can be correlated to the actual events that cause the corrosion. The use of more sophisticated instrumentation and data loggers has allowed the users to "look inside of their pipes."

With the newer instrumentation and software a whole range of innovative non-intrusive internal corrosion monitoring devices are now on the marketplace.

Major corrosion monitoring techniques

The major corrosion monitoring techniques are:

1. Coupons (C)
2. Electrical resistance (ER)
3. Linear polarization resistance (LPR)
4. Galvanic (ZRA)
5. Hydrogen probes and patches (H_2)

Several newer corrosion monitoring techniques are being tested in the laboratory, and in the field, and may prove of interest in the future. They are:

1. Hydrogen patches (H_2)
2. Impedance (EIS)
3. Electrochemical noise (ECN)
4. Surface resistance readings
5. Others (i.e., radiography, acoustic emission, etc.)

Weight loss corrosion coupons

Weight loss corrosion coupons are probably the oldest and still most widely used method of corrosion monitoring. Coupons are simply a specimen of metal that is firstly weighed, then exposed into a specific environment, removed, and then cleaned and reweighed to determine the amount of weight loss corrosion that has occurred over a specific period of time. NACE has a standard method of determining the weight loss, and the result of this determination gives one the amount of weight loss in mils per year (MPY), which is the most widely used reference. This reference is also used on most other corrosion monitoring methods. MM per year is also used, as are several different measurements, but, MPY is by far the most widely used.

Weight loss corrosion coupons are available in many different shapes and materials. The size is not that important once one realizes that the more surface area exposed to the environment, the more accurate will be the readings, and the more quickly a reading can be obtained. The most common shape is commonly called a strip coupon and is most often available in the following sizes; $\frac{1}{2}' \times \frac{1}{16}' \times 3'$ or $\frac{7}{8}' \times \frac{1}{8}' \times 3'$. Other configurations are also used such as rod or pencil coupons, pre-stressed coupons of several shapes, and "lifesaver" shaped coupons that are often fitted with electronic probes.

Weight loss corrosion coupons are available in any material desired. The material is, however, most often either that of the material containing the environment (i.e., the pipe, vessel, or tank) or a low carbon steel that is very susceptible to corrosion. If you are trying to duplicate the corrosion that is occurring on the pipe, you should use coupons of the same material. If you are using the coupons to determine if the inhibitor is filming and remaining in place, a low carbon steel coupon can be used. However, in using low carbon coupons, it must be remembered that if you can stop corrosion on these coupons, the corrosion rates of the pipe itself will not necessarily be affected.

A third use of coupons is for material selection. If a pipe, vessel, or tank is going to be replaced it is often advisable to install coupons of several different alloys to check the performance of each. This allows the engineer to verify his material selection of the new pipe or place on order the material best suited for the particular environment with which he may be dealing.

Coupons should always be used to both verify the probe readings and for long-term verification of corrosion rates.

The advantage of coupons is that they are relatively inexpensive, the weight loss result is positive, samples of corrosion product or bacteria can be obtained from the surface for further observation and testing, and coupons are not subject to instrument failure like the electronic methods of corrosion monitoring. The disadvantage of corrosion coupons is that the results take a long time to obtain and coupons can only give average readings. Coupons will tell you the corrosion averages from when the coupon was installed until when it was removed. Coupons do not, however, take into account that corrosion is not usually constant.

Electrical resistance probes and instruments

Electrical resistance probes and their associated instruments measure the resistance through a sensing element that is exposed to the environment to be studied.

The principle of electrical resistance probes and instruments is that electrical resistance of the sensor, having a fixed mass and shape, will vary according to its cross sectional area. As corrosion and/or erosion occurs, the cross sectional area of the element is reduced, thus changing the resistance reading. This change in resistance is compared to the resistance of a check element that is not exposed to weight loss corrosion or erosion, and if the two resistance readings are expressed as a ratio, then changes in this ratio can be shown as a corrosion rate.

Probes are constructed in various designs and materials depending upon the pressure, temperature, velocity, and other process parameters of the system to be monitored.

Instrument availability falls into many categories, but the general configurations of most electrical resistance instruments are as follows:

- Portable: Portable instruments allow the operator to take manual readings in the field. Newer portable instruments also have the ability to record in the memory facilities the measurements made, tag numbers, and often other data as well. In addition, readings can be observed in the field as they are taken or can be left to be uploaded onto a PC for recording or charting at a later time.

- Data collection instruments: Data collection units can be used anywhere and are especially valuable for use on unmanned or remote sites and locations with difficult access. Measurements are made automatically at the probe as often as required and this data is stored onboard for later retrieval when it is either economically feasible or convenient for the operator. Some data collection units can be programmed in the field or from a PC in the office.
- Transmitter base instruments: In these cases the signal from the probe is transmitted to either a standalone instrument or PC or incorporated into a full operational SCADA or instrument package for either a full time, scanned time, or alarm function readout. The limitations are dependent on the instrumentation package available in each facility.

Electrical resistance monitoring can be used in almost any environment including a "dry" system. Electrical resistance monitoring will also measure erosion. The measurements one gets when using electrical resistance technology are averages over time, like coupons, and these readings average the corrosion rate between readings. However, it should be pointed out that the newer instrumentation allows so many readings to be taken in such short times that the averaging time is cut to almost nothing, thus approaching real-time corrosion monitoring. The only drawback to this technique is that in systems where conductive depositions are being formed, the deposition may interfere with the resistance readings. Also, as with all sensors, they only measure corrosion when the sensing element corrodes; thus they have a life expectancy that is corrosion dependent. Simply put, probes must be periodically replaced.

Linear polarization resistance probes and instruments

LPR probes and instruments measure the ratio of voltage to current, the polarization resistance, by applying a small voltage, usually between 10 and 30 millivolts, to a corroding metal electrode and measuring the corrosion current flowing between the anodic and cathodic half cells or electrodes. The polarization resistance is inversely proportional to the corrosion rate.

LPR probes also come in a wide variety of configurations and materials. Two- and three-electrode probes are available, the principle difference being that the three-electrode probe attempts to minimize the solution resistance by introducing the third electrode, or reference electrode, adjacent to the test electrode, in order to monitor potential in the solution with a view to reducing the relative contribution of the solution resistance to the series resistance path.

The sensing elements, or electrodes, are made from any materials and in many configurations, usually rod shaped electrodes or flush electrodes. The same considerations in choosing the material must be made as in selecting coupons or electrical resistance probes.

Probes are constructed in various designs and materials depending upon the pressure, temperature, velocity, and other process parameters of the system to be monitored.

Instruments are available in portable, data logger, or rack-mounted design depending upon the budget and technical need. As LPR gives a real-time result that is continuous, it is much more important to use instrumentation connected to the probe full time, as readings will vary according to the corrosivity of the liquid measured. LPR probes do require submersion into an electrolyte and will not function in "dry" environments.

LPR measurements are much more predominantly used in water side corrosion environments where fast readings are required, so that chemical corrosion inhibitor being injected can be tuned to prevent the corrosion from taking place or at least kept to a minimum.

Galvanic probes and instruments

Galvanic probes and their related instrumentation expose two dissimilar elements into an electrolyte. These elements of electrodes are attached through an ammeter and the resulting readings offer insight into the corrosion potential of the fluid. It should be noted that worldwide galvanic probe corrosion monitoring is not the most widely used method and although these techniques have their strong proponents, many do not use this technique.

Very much like a coupon, galvanic probes come in many different configurations and materials. The most common configuration of the sensing element is rod type electrodes. Flush elements also are available in several different configurations. Element choice is dependent upon the service and the expected degree of corrosion. Generally speaking, the more sensitive the sensing element, the shorter the life, and conversely, the less sensitive the sensing element, the longer the life. Depending upon the service and the expected corrosion rates, the proper element shape and sensitivity can be specified.

The sensing elements of galvanic probes are available in any metal material and the same consideration in choosing a material for a coupon does not have to be considered when choosing the material for the galvanic probe sensors or electrodes. Generally speaking, the two electrodes must be of dissimilar material. This dissimilarity produces a natural current flow through the electrolyte. Measuring this current flow provides the data to be studied.

One of the more frequent uses of a galvanic probe is for oxygen detection. In this case a brass cathode electrode and a steel anodic electrode are used and changes in the level of dissolved oxygen in treated water will produce readings. Often galvanic probes are installed downstream of water pumps and if the seals start to leak, the pump's air (oxygen) will be sucked into the system and the probe will be alerted to this condition.

Hydrogen permeation technology

Many advances have been made since the original concept of the hydrogen probes first used in the 1930s and they will be covered in this section. The two basic categories of hydrogen permeation monitoring devices are those which use a sensor to measure the H_2 resulting from atomic hydrogen permeation and those which actually measure the H_2 resulting from atomic hydrogen permeation through the wall of the pipe, vessel, or tank.

Hydrogen permeation or hydrogen flux permeation usually originates from the hydrogen atoms that are liberated at the cathode during the electrochemical process of corrosion. It is also possible that the hydrogen atoms may exist due to a chemical reaction inside the vessel or pipe unrelated to corrosion processes. A third possibility is that hydrogen atoms can exist in the steel due to the manufacturing and/or welding procedures. Hydrogen "bake-out" procedures can remove the hydrogen that is present during manufacturing or welding and should be used. All hydrogen permeation devices use hydrogen permeation, or hydrogen flux permeation, and the resultant H_2 accumulation is the method of monitoring the rate of this permeation.

Not all of the hydrogen atoms generated by a corrosion reaction will necessarily migrate into the steel of the pipe, vessel, or tank. Depending upon the process, and the corrosion environment, a certain percentage of these atoms can recombine on the inside surface of the pipe to form molecular hydrogen gas (H_2). This hydrogen gas ultimately goes into the product stream and is lost in the process flow as it is carried down the line. Poisons in the system play a very large role in how much of the atomic hydrogen that is liberated actually goes into the steel itself.

Atomic hydrogen migrates along grain boundaries, and therefore, not all atomic hydrogen goes in a straight line to the outside of the pipe or vessel. There are theories that some hydrogen atoms can migrate along the wall of a pipe before exiting to the outside wall but, regardless of this theory, we can assume the majority of the atomic hydrogen goes relatively straight through the wall. Once through the wall of either the sensor, the pipe, the vessel, or the tank, one hydrogen atom is naturally attracted to another, forming H_2, which either forms inside the annulus of the sensor, escapes into the patch type device or goes off into the air.

Insert type hydrogen probes expose a corroding element to the environment and as atomic hydrogen permeates the steel and reaches a cavity, the hydrogen atoms combine to form H_2, which builds up pressure in the cavity and this pressure build up is the basis for the data that may be, and probably is, related to internal corrosion. External H_2 pressure patches also use this same principle.

Another type of hydrogen patch probe is the electrochemical type. This is available as either a permanent patch or a small temporary designed patch that gives a quick reading. This type of device works by polarizing a palladium foil held to the wall of the metal by a transfer medium such as wax, and as the palladium foil is polarized it acts as a working electrode, quantitatively oxidizing the hydrogen as it emerges from the wax. The current induced by the instrument is directly equivalent to the hydrogen penetration rate.

A third type of hydrogen patch monitoring device uses a thin plate designed to capture the molecular hydrogen as it is generated from the atomic hydrogen as it escapes to the outside surface of the pipeline and into the annulus between the outside of the pipe and the underside of the foil.

Since the definable chamber of this H_2 plate is under vacuum, these hydrogen atoms react almost instantly to form molecular hydrogen gas (H_2) once they enter the annulus. H_2 molecules are many times larger than the single hydrogen atom, and are unable to escape from the vacuum chamber by either going back into the pipe or through the foil itself. As more and more atomic hydrogen escapes into the annulus and the build-up of H_2 takes place, the vacuum will then begin to decay or decline in a ratio proportional to the intensity of the atomic hydrogen permeation taking place from the inside surface of the pipe. Use of a vacuum is very important, as conventional pressure build-up hydrogen probes are very susceptible to temperature changes, whereas a vacuum or partial vacuum is relatively immune to temperature changes. Therefore, the data is much more even and not subject to the wide swings that pressure build-up devices experience due to these swings.

Regardless of the cause of atomic hydrogen permeation, the result of this permeation can cause the following to occur:

1. Hydrogen blisters
2. Hydrogen induced cracking (HIC)
3. Hydrogen embrittlement
4. Sulfide stress cracking
5. Carbide phase attack

None of the above conditions are desirable in pressure containing devices that may contain volatile or polluting fluids or gases and all of the above can result in sudden catastrophic failures.

In some refinery and petrochemical situations hydrogen flux flow is extremely high and no appreciable weight loss corrosion is detectable. However, regardless of the source of free atomic hydrogen migrating along the grain boundaries, this flow of atomic hydrogen can cause long-term and/or catastrophic failures.

The quantity and intensity of hydrogen flux are determined by the following factors:

1. Amount of H_2S in the system
2. System poisons (cathodic reaction poisons)
3. Type of corrosion process occurring

4. Type of material of the pipe or vessel
5. Method of material construction
6. Deposit composition

Generally speaking, if the environment in the system has amounts of H_2S, or if it contains cathodic poisons, more of the atomic hydrogen being generated at the cathode will be driven through the wall of the pipe. Although the intensity of the corrosion reaction may vary from system to system, changes in intensity within a given system can usually be directly correlated to the corrosion process. In other words, if a hydrogen flux rate causing a decay in vacuum of 10 kPa/day is reduced to 5 kPa/day, the corrosion on the inside may be cut in half if all other conditions remain the same.

It is this corrosion intensity relationship that allows the user to experiment with inhibitor corrosion control programs or process changes and to be able to see within minutes whether these changes are beneficial, or whether additional changes are required to stop, or at least lower, corrosion. Similarly, changes in operation parameters can also be observed rapidly.

This newly improved method of near real-time monitoring does not conflict with any existing internal corrosion monitoring devices, but rather adds a new dimension. For example, there is nothing wrong with knowing that there is a general corrosiveness to water and that this corrosiveness increases and decreases. However, the point is, does this manifest itself in actual system corrosion, or are the system corrosion processes driven by factors largely independent of water chemistry?

SUMMARY AND CONCLUSIONS

The development of the many different corrosion monitoring devices and their modification and improvement over time has allowed the pipeline industry to be able to monitor accurately and rapidly almost all of the metal pipes and structures desired. The speed and control of the newest generation of these different techniques has been found to exceed all expectations with changes in corrosion intensity being measured in minutes and hours rather than what the previous method offered in days and weeks or even months.

The recent addition of automated systems allowing the data to be transmitted to data loggers, PCs, or SCADA type systems allows for the most modern up-to-date applications. By being able to assimilate this data in a very quick manner, the corrosion events can be compared to other operational parameters allowing the operator to fine-tune these other parameters and thus lower or stop the corrosion.

The advantages of internal corrosion monitoring

The advantages of internal corrosion monitoring include the following:

1. Evaluation and fine-tuning of inhibitor programs.
2. Evaluation and scheduling of pigging and other inspection programs.
3. Evaluation of process changes and upsets.
4. Optimization of the corrosion resistance of an operation.

5. Life prediction study verifications.
6. Insurance rate adjustment possibilities.
7. Management and operation awareness of corrosion or success in corrosion mitigation.
8. Compliance with federal, state, local, and industry rules, practices, and guidelines.

Internal corrosion monitoring during pigging operations

The above listed monitoring techniques are grouped into either intrusive or non-intrusive methods. Intrusive methods usually involve inserting a coupon or probe into the pipeline or piping system at pump or compression stations, pig traps, drip pots, or other such related piping arrangements.

In order to accomplish this insertion it is necessary to operate either at full operating pressure or to shut down the pipeline for insertion and removal of the monitoring sensor. Another alternative is to place the sensors on a by-pass with a block-and-bleed capability, but many feel that the environment of corrosion, and hence the data received from by-pass, can be different from the actual flowing line itself. For this reason most operators do not use by-passes unless there is no alternative.

Over the past several years with the use of pigging it has become fairly common practice to use flush mounted coupons and probes. With the flush mounted probes, there is no interference with the passage of the pig. Using standard access fitting installed at any point around the circumference of the pipe that is suitable for the probe or coupon to be in a corrosive environment, an access fitting can be installed. With proper calculations such as the pipe wall thickness, weld gap and design of both the sensor, sensor holder and plug within the access fitting, the sensors can be truly flush with the inside of the pipe. In addition to allowing the passage of pigs, this design will place the sensor at the pipe wall where the corrosion is occurring. The potential for corrosion near the middle of the gas or liquid stream has little or no bearing on the general corrosion occurring on the pipe wall itself.

Non-intrusive methods do not interfere in any way with pigging operations. Unfortunately, at this point in time, only a few non-intrusive methods are field proven and only a few give the operator either enough or complete internal corrosion monitoring data. While hydrogen patch devices, for instance, work well in many environments, there are many applications where they do not seem to work at all.

Stress corrosion cracking (SCC) threat

SCC propensity

There are two types of SCC normally found on pipelines:

(1) High pH (pH 9 to 13)
(2) Near-neutral pH SCC (pH 5 to 7)

The High pH SCC caused numerous failures in the USA in the early 1960s and 1970s, whereas near-neutral pH SCC failures were recorded in Canada during the mid 1980s to early 1990s. The SCC failures have been reported from areas including Australia, Russia, Saudi Arabia, South America, and other parts of the world.

SCC defined

SCC is a brittle failure mode in otherwise ductile material. This unexpected and sudden failure of ductile metal is also under a tensile stress in a corrosive environment. This condition is further aggravated if the metal is at elevated temperature. For pipeline this would be a temperature above 40°C (104°F).

SCC is an anodic process; this is verified by the application of CP, which is used as a remedial measure. Usually there is some incubation period for the cracks to be detected; during this period the cracks originate at a icroscopic level. This is followed by active progression of cracks. Sometimes cracks may be self-arresting, for multi-branched transcrystalline SCC, because of localized mechanical relief of stresses.

If any one of the two factors (stress and specific environment) responsible for SCC is removed, the SCC will not occur and further progression of cracks may stop. SCC is associated with little general corrosion. In fact, if extensive general corrosion is present, SCC is less likely to occur. However, trapped corrosion products may initiate another SCC cell under the general corrosion.

There is general understanding that steel with tensile strength above 130 ksi (896 MPa) is susceptible to SCC, which is a true statement. However, it does not mean that steels below that level of tensile strength are not susceptible to SCC. As mentioned above, stress and the specific environment are the main contributors to the cause of SCC and temperatures above 40°C (104°F) add to that condition.

Factors essential to cause SCC

Environment

SCC is highly chemical specific; certain materials are likely to undergo SCC only when exposed to a small number of chemical environments. The specific environment is of crucial importance, and only very small concentrations of certain highly active chemicals are needed to initiate SCC. Often the chemical environments that cause SCC for a given material are only mildly corrosive to the material in other circumstances. As a result of this phenomenon, the parts with severe SCC may appear unaffected on casual inspection, while they might in fact be filled with microscopic cracks. Unless a conscious effort is made by a specific targeted inspection plan to detect SCC, this special condition can mask the presence of SCC cracks for a long time. SCC often progresses rapidly, leading to catastrophic failures.

The second issue with SCC is the stresses. Stresses can be the result of the crevice loads due to stress concentration, or can be caused by the type of assembly or residual stresses from fabrication (e.g., cold working); the residual stresses caused by fabrication can be relieved by annealing.

Environment Figure 1-5-3 is a critical causal factor in SCC. High-pH SCC failures of underground pipelines have occurred in a wide variety of soils, covering a range of colors, textures, and pHs. No single characteristic has been

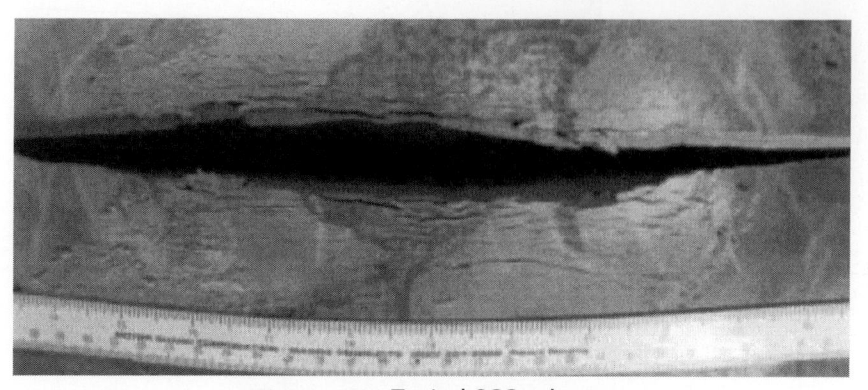

Figure 1-5-3 Typical SCC colony.

found to be common to all of the soil samples. Similarly, the compositions of the water extracts from the soils have not shown any more consistency than the physical descriptions of the soils. On several occasions, small quantities of electrolytes have been obtained from beneath disbonded coatings near locations where stress corrosion cracks were detected. The principle components of the electrolytes were carbonate and bicarbonate ions and it is now recognized that a concentrated carbonate–bicarbonate environment is responsible for this form of cracking. Much of this early research focused on the anions present in the soils and electrolytes. In addition, the coating failure, the local soil, temperature, water availability, and bacterial activity have a critical impact on SCC susceptibility. Coating types such as coal tar, asphalt, and polyethylene tapes have demonstrated susceptibility to SCC. High efficiency coating systems like 3LPE and FBE have not shown susceptibility to SCC.

Loading

Loading is the second most important parameter to contribute to SCC. Cyclic loading is considered a very important factor. The crack tip strain rate defines the extent of corrosion or hydrogen ingress into the material. There has been no systematic effect of yield strength on SCC susceptibility.

Other factors

Certain types of welds, especially low frequency welded electric resistance welding (ERW) pipe, have been found to be systematically susceptible to stress corrosion cracking (SCC). Non-metallic inclusions have also had limited correlation to SCC initiation.

High pH SCC

High pH SCC can be called a classical SCC. The phenomenon was initially noted in gas transmission pipelines. In practical terms it is often found within 20 kilometers (about 12.5 miles) downstream of the compressor station. High pH SCC normally occurs in a relatively narrow cathodic potential range (-600 to -750 mV $Cu/CuSO_4$) in the presence of a carbonate and bicarbonate environment in a window from pH 9 to pH 13. System temperature should be greater than 40°C (104°F) for high pH SCC possibility. The crack growth rates decrease exponentially with lower temperature.

Intergranular cracking mode generally represents high pH SCC. A thin oxide layer is formed in the concentrated carbonate–bicarbonate environment, which surrounds the crack surfaces and provides protection. However, due to changes in loading or cyclic loading, there is crack tip strain resulting in breakage of oxide film. This results in crack extension due to corrosion. Because of such a stringent environmental requirement for high pH SCC initiation, this type of SCC is not as prevalent as the near-neutral pH environment SCC. The high pH SCC has been primarily noted in gas transmission lines associated with higher (greater than 40°C) temperature.

High pH SCC: Integrity management strategy

To evaluate and establish the extent of SCC susceptibility, the following steps must be taken:

a. Evaluate the selection of material, coating, and other operational conditions that are conducive for SCC.
b. Review the ditch coatings survey to identify locations of holiday and match them with high stress levels. High stress level is defined as stress equal to or exceeding 60% stress of the specified minimum yield strength (SMYS) of material.
c. Identify and match the stress with high temperature locations.
d. Match the inspection report and identification of coating failures with corrosion, even minor corrosion, to identify the potential for SCC.
e. Excavate to identify susceptibility, conduct MPI on suspected locations to locate SCC. Meet mandatory requirements of due diligence inspection.

Near-neutral pH SCC

The near-neutral pH SCC is a trans-granular cracking mode. The phenomenon was initially identified in Alberta, Canada, and has been followed by reports from pipeline operators in the USA. The primary environment responsible for near-neutral pH SCC is the diluted groundwater containing dissolved CO_2 gas. As with high pH, the CO_2 is generated from the decay of organic matter. Cracking is further exacerbated by the presence of sulfate reducing bacteria (SRB). This occurs primarily at the sites of disbonded coatings, which shields the cathodic current reaching the pipe surface. This creates a free corrosion condition underneath the coating, resulting in an environment with a pH of around 5 to 7.

A cyclical load is critical for crack initiation and growth. There are field data that indicate that with a decreasing stress ratio there is an increased propensity for cracking. Hydrogen is considered a key player in this SCC mechanism, where it reduces the cohesive strength at the crack tip. Attempts have been made to relate soil and drainage type to SCC susceptibility; however, limited correlations have been established.

There has been no correlation to a clear threshold for SCC initiation or growth. The morphology of the cracks is wide, with evidence of substantial corrosion on the crack side of the pipe wall.

Near-neutral pH SCC: Integrity management strategy

To evaluate and establish the extent of SCC susceptibility, the following steps should be taken:

a. Evaluate the material selection and coating system to ensure they are compatible with the SCC conditions.

b. Review and analyze the corrosion inspection survey reports to identify areas of linear corrosion or small pitting corrosion locations to identify sites for SCC susceptibility.
c. Identify and analyze locations of high cyclical pressure combined with a high operating pressure.
d. Conduct bell-hole inspection, excavate at several of these locations to develop extent of SCC on the pipeline system.
e. Conduct MPI to identify presence of cracks.

Additional parameters such as soil and drainage should also be considered for SCC susceptibility; however, the pitfalls of this step must be born in mind, as both very poor and well-drained soils have shown susceptibility to SCC.

Threats from Table 1-5-1 manufacturing defects

This is one of the static threats of the pipeline. This involves both the inherent defects in pipe and weld. For new construction and new pipes the process should begin with the selection of a good quality steel source and pipe-making practices. The selection must consider all aspects of corrosion and stresses that the life of the pipeline is expected to encounter. The pipe material selection process for new pipe segments should, as a minimum, include the following:

a. Evaluation of steel production process.
b. Segregation and chemical composition control during production process.

Table 1-5-1 Perspective Integrity Management Plan for Time Dependent Threats

Inspection technique	Maximum time intervals	Operating stress criteria		
		At above 50% SMYS	*At or above 30% and up to 50% SMYS*	*Less than 30% SMYS*
Hydrostatic test	5	Test pressure to 1.25 MAOP	Test pressure to 1.4 MAOP	Test pressure to 1.7 MAOP
	10	Test pressure to 1.39 MAOP	Test pressure to 1.7 MAOP	Test pressure to 2.2 MAOP
	15	Not permitted	Test pressure to 2 MAOP	Test pressure to 2.8 MAOP
	20	Not permitted	Not permitted	Test pressure to 3.3 MAOP
Inline inspection	5	Predicted failure (P*f*) above 1.25 MAOP	Predicted failure (P*f*) above 1.4 MAOP	Predicted failure (P*f*) above 1.7 MAOP
	10	Predicted failure (P*f*) above 1.25 MAOP	Predicted failure (P*f*) above 1.7 MAOP	Predicted failure (P*f*) above 2.2 MAOP
	15	Not permitted	Predicted failure (P*f*) above 2 MAOP	Predicted failure (P*f*) above 2.8 MAOP
	20	Not permitted	Not permitted	Predicted failure (P*f*) above 3.3 MAOP
Direct assessment	5	Sample of indications examined	Sample of indications examined	Sample of indications examined
	10	All indications examined	Sample of indications examined	Sample of indications examined
	15	Not permitted	All indications examined	All indications examined
	20	Not permitted	Not permitted	All indications examined

c. Slab cleaning and scarfing to remove surface defects like scabs, laps, slivers, and remove inclusion to eliminate lamination, in the final rolled plates and coils.
d. Control of rolling process, with automated temperature control for rolling operation.
e. Online-automated X-ray thickness scan and control on the finishing rollers to control uniform thickness on plate/coil material.
f. Evaluation of the pipe mill, and their manufacturing procedure specification (MPS).
g. Selection of minimum residual stress in pipe rolling process.
h. Establishment of inspection and testing regime and acceptance criteria.
i. Control of transportation and storage of pipes, until they are welded, coated, and buried.

The API 5L specification requirements serve as a basis for all control of chemical composition, the dimensions, and mechanical tests. Further additions and changes may be made as necessary upon evaluation of specific design requirements.

For all pipelines, new or old, the following data must be collected as a minimum for risk assessment:

- Pipe material source
- Manufacturing process
- Pipe material grade
- Type of weld seam
- Joint factor

- Operating pressure temperature
- In-service history of pipe segment

The pipes manufactured by welding processes such as butt welding, lap welding, hammer welding, furnace welding, flash welding, and low frequency electric resistance welding (ERW) are all susceptible to failure. All pipes that have a joint factor of less than 1.0 have inherent manufacturing threats.

Additionally Table 1-5-2, conditions such as the following must be considered, and if evidence suggests, such threats

Table 1-5-2 Longitudinal Weld Joint Factors

Specification	Weld joint type/pipe class	Joint factor
ASTM A 53	Seamless and electric arc welded (ERW)	1
	Furnace butt welded: continuous welded	0.60
ASTM A 106	Seamless	1
ASTM A 134	Electric fusion arc welded	0.8
ASTM A 135	Electric arc welded	1
ASTM A 139	Electric fusion welded	0.8
ASTM A 211	Spiral welded steel pipe	0.8
ASTM A 333	Seamless	1
ASTM A 381	Double submerged arc welded	1
ASTM A 671	Electric fusion welded Class 13, 23,33,43, and 53	0.8
	Electric fusion welded Class 12, 22,32,42, and 52	1
ASTM A 672	Electric fusion welded Class 13, 23,33,43, and 53	0.8
	Electric fusion welded Class 12, 22,32,42, and 52	1
API 5L	Seamless	1
	Electric resistance welded	1
	Electric flash welded	1
	Submerged arc welded	1
	Furnace butt welded	0.6

must be considered in depth for conducting the risk assessment:

- 50-year-old pipe system
- Cast iron pipes (special attention should be given in seismic and landslide areas)
- Mechanically coupled pipes
- Oxy-acetylene welded pipes
- Low temperature environments
- Pipes exposed to land movements (earthquakes and landslides)

Risk assessment and mitigation

Validation of weld seam for steel pipes must be done through hydro-test to at least 1.25 times the MOAP. Raising the MOAP or changing the operating pressure must be conducted based on this assessment. The steel pipes that fail the hydro-test must be removed and replaced with suitable pipeline.

The risk mitigation for cast iron pipes may be achieved by either replacing the pipe or by lowering the stress level.

Construction and fabrication related threats

The construction threats emanate from activities like welding of girth-welds, fabrication of bends that may have wrinkled and buckled surfaces, and in threaded pipes it comes from the stripped threads, broken pipe, etc.

Sometimes the presence of some of these construction and fabrication threats may not be detrimental by themselves;

however, their existence in the presence of other threats can be a cause of compromised integrity. The risk evaluation of such threats must be carried out in that context, by data integration and examination of all aspects of threat.

Identification of these threats is the first step toward the management of the associated risks. Once positively identified, the information must be collected from a pipeline segment to initiate risk assessment. Relevant data include the following:

1. Identification of the wrinkled bend
2. Identification of failed coupling
3. Review of bend making procedure
4. Review of welding procedures and qualification records
5. Review of NDT information on welds, establish traceability of each joint and repair, etc.
6. Pipeline inspection reports, streamline reports
7. Any reinforcement provided after construction
8. Hydro-test information
9. Any possible outside damage
10. Depth of cover and properties of soil
11. Pressure and temperature range (designed versus operational)
12. Operating pressure and temperature history
13. Identification of any possible fatigue mechanism

The review of welding procedure and qualification records is done to ensure that the resulting weld is compatible with properties of the parent metal. The visual inspection of wrinkles and damaged threads on couplings is done to ensure that the system is integrally safe; if not, these are to be replaced. Any movement of soil must be ascertained to ensure there is no additional lateral or axial stress to the pipe system.

On assessment of these threats and establishment of a suitable integrity program, the effectiveness of the program can be measured by the number of leaks occurring due to construction and the number of repairs or corrective steps initiated to maintain the integrity of system.

Equipment failure threat

The term equipment in this context is used to identify parts used in the pipeline system other than the pipe and fittings, such as gasket seals relief, O-rings, pump, etc., that are often the cause of failure and sometimes, a small leak or failure from the equipment can lead to major failures and accidents.

Industry experience has shown that gaskets, O-rings, control and relief valves, seal and pump packing have been causes of incidents.

The information and data collected and reviewed for risk assessment from the failure of equipment will include the following:

1. Years of service of the equipment
2. Any previous failure

3. Relief valve settings and past activation of relief valve
4. Flange gasket type and any past failures
5. Regulator set point compared with the tolerance
6. Past O-ring failure, ring material, and service environment compatibility
7. Seal packing information on any history of failure

The operations and maintenance procedure of the plant should address periodic inspection of these pieces of equipment and replace them with adequate frequency.

The risk assessment and performance discussed above can be judged by the number of failures in a given space of time. Often this is established by determining the mean time between failures (MTBF). MTBF is the arithmetical mean ($1/\lambda$) of the time between number of failures of a single system or component, or all failures of a population of similar systems or components.

There are a variety of ways MTBF can be calculated, often driven by the availability of information. Often, it is the time between successive failures or the time from the end of the last repair to the next failure. It can also be calculated as the time the component has been in service.

$$\mathrm{MTBF} = (1/\lambda)$$

where, λ is rate of failure.

The best approach is the one that is consistent and has uniformity of application to the given system.

Third party damage threat

This is one of the time dependent threats; even if there is no damage in inspection it can happen any time. Due to the uncertainty of its occurrence, an effective threat mitigation program is required. Shallow depth of buried pipe in agricultural land is especially susceptible to third party damage.

Pipeline ROW and Access

- **The pipeline must be accessible.**
- **Any erosion damage must be corrected.**
- **Any vandalism must be repaired.**
- **Pump station, valve stations and other assets must be secured.**
- **Monitoring must be proactive.**

Pipeline ROW Erosion - 1

Pipeline Vandalism

A pipeline, if it is maintained properly and has good integrity, can perform its given function and operate safely and efficiently. This gives the operators, regulators, owners, and the general public confidence in its safe operation.

Third party damage is the damage caused to the pipeline system by people or activities that are not in any way responsible for maintenance or operations of the pipeline system. This could be due to vandalism or a case of people performing some work near the pipeline that is not related to the pipeline and accidently damaging it. Control on land encroachment and monitoring the length of pipeline should be carried out on a regular basis. Several modern steps have been taken and tools are available for monitoring, including on-line monitoring and arial observation, coupled with GPS coordinates to locate potential sources of damage encroachment and unplanned activities around the pipeline, allowing immediate reaction.

The data to be collected for risk assessment should include the following:

a. Past history of vandalism to the pipe and also to other pipelines in the area
b. Bell-hole inspection data of the pipe location hit

c. Any history of leaks due to damage and their location
d. ILI inspection reports of dents and gouges in the top half of the pipe
e. One-call records
f. Encroachment records

The risk assessment should establish the possible level of threat and plans must be put in place to address the failures, which can sometimes be high consequence and constitute an emergency situation.

Prevention is the best step to control third party damage threats to pipeline. Prevention measures are the first line of defense from third party damages. However, if damage does occur, repair is the next step.

Threat from incorrect operations

If operating procedure is not followed, or the limits of operating conditions are exceeded, pipeline can fail. Such excesses in operating conditions pose a threat to the pipeline system. This may include exceeding the pressure or temperature limits.

To support risk assessment, the following data must be collected and analyzed:

a. Review of established procedure
b. Past audit reports and recommendations
c. History of past failures involving operation procedure
d. Qualifications of operators

If the review indicates a flaw in the operating procedure then steps must be initiated for their immediate correction.

Weather related threat

The threats associated with weather and outside force are defined as lightning excessive cold and snow, heavy rains, landslides, floods and so on, that are beyond any prediction or control.

Some of these data can be obtained from local land and weather management bodies. Historical highs and lows of any area can be obtained from published records for several years and at the design stage these data are reviewed and considered. Yet, sometimes weather and natural calamity records are incomplete.

For such unknown threats, as much detail as possible should be gathered. This involves collection of most of the data we have discussed in relation to other threats, and more. Thus the data to be collected for risk assessment can vary; however, the following minimum should be collected:

a. Pipe grade
b. Pipe diameter
c. Pipe wall thickness
d. Design and operating stress
e. Type of weld joint and details
f. Land topography
g. Soil condition
h. Water crossing

i. Wetlands
j. Soil liquefaction susceptibility
k. Earthquake faults
l. Profile of ground acceleration (greater than 0.2 g)
m. Depth of frost line

The risk assessment criteria should evaluate all of the above data and conditions. Where such risk is identified, the operations and maintenance procedure should be suitably corrected and inspection types and frequencies increased.

HANDLING UNCERTAINTY

Definition of uncertainty

- **Limitation of the measurements and tools, personal performance, and other unknown factors.**
- **Example: Sabotage can bring about an unexpected risk that causes failure.**

Now we know the threats and understand the mechanisms underlying those threats. We know how to assess the risk associated with each of these known threats and the options in mitigating the threats. However, while addressing known threats, we must also be aware of possible uncertainty involved, as we can't know and control enough of the details to eliminate risk.

At any given time and location there are thousands of forces acting on a pipeline, the magnitude of which are literally unknown and at times unknowable. A reference to chaos theory and entropy would not be out of place here.

Initially it is important to decide on how to deal with these uncertainties in assessing risks. The best way is to use a process of elimination by treating each possibility as potential and assuming the worst until data shows otherwise. In risk assessment there is always a 50% possibility of error.

Risk assessment could be either: (1) called good when it is actually bad; or (2) called bad when it is actually good. In the first case, since it is called good, no flag will be raised and it will not be further investigated. The error will persist, as it will not be found out until an incident occurs or a fresh mind such as an auditor investigates it. In the second case, since a flag is raised, it will be investigated and the error will be found and corrected. This can often lead to overreaction and distrust in the system is induced, resulting in extra resources being spent and some good pipelines being penalized by a poor assessment process.

It is also important that a risk assessment identifies the role of uncertainty in its use of inspection data. Information has a finite life span; as time passes, the utility of available information and data diminishes. This is true for all data obtained from inspection and surveys.

ROLE OF INSPECTION AND TESTING

Inspection and testing play a significant role in assessing pipeline failure potential. Several sets of data for pressure test ILI and non-destructive examination (NDE) inspection are collected over a period of time for a section of pipeline. The role of a good integrity management system is to present this data as useful information.

The integrity management models clearly separate the information obtained by inspection from that which is derived by analysis. For example, consider the inspection data and the analysis and evaluation that give the rate of damage; both are very distinct and separate pieces of information, though the analysis is dependent on the accuracy of the data obtained from inspection.

As discussed earlier, the risk is the mathematical product of probability and the consequence of failure, so it is logical that the second part of the risk assessment has to be the understanding and analysis of consequences. This is the determination and study of the result of failure that may occur due to the risks assessed. Reduction in risk level can reduce the consequences of the failure. Adjustment to the safety factors and design criteria can achieve the goal of reducing the risk and resulting consequence level.

Determining the level of consequence of any failure involves determining the potential area of impact of an event. The impact area is a function of pipe diameter and the system pressure. The calculated area of impact is reported as the radius of impact. In US customary units, the relationship can be expressed as follows, where the constant 0.69 is the factor for natural gases.

$$\text{Radius of impact in ft (r)} = 0.69\,(d \times \sqrt{p})$$

where d is the outside diameter of the pipe in inches and p is the pipeline segment's MAOP in psig.

Other variables such as depth of the buried pipe should also be considered. The area of impact is the area covered between two adjacent circles whose radius is calculated by the above formula. The number of dwellings within the area of impact is used to determine the consequence area. While the properties of gas, such as richness, toxicity, and other harmful effects are a major factor in determining the full consequences of an incident, the following general factors must be considered for all consequence determination:

a. Population density, especially high population areas, defined by the Census Bureau as urbanized areas. Other populated areas, defined by the Census Bureau as places that contain a concentrated population.
b. Unusually sensitive areas.
c. Commercially navigable waterways.
d. Damage to property.
e. Damage to environment.
f. Proximity of population to the pipeline.
g. Damage to public convenience.
h. Proximity to hospitals, schools, and other sections of the population with limited mobility.
i. Potential for secondary failures.

j. Effects of unignited gas.
k. Effect on the supply of gas to essential services and the general population.

In addition to the above, the term "unusually sensitive areas" is defined as drinking water or ecological resource areas that are especially responsive to environmental damage from a hazardous liquid pipeline release. The federal government in the USA has applied this definition to identify high consequence areas (HCAs) and has made maps depicting these locations for pipeline operators. Operators are also responsible for independently evaluating information about the area around their pipeline to determine whether a pipeline accident could affect a nearby, but not adjacent, HCA.

MANAGING POTENTIAL RISKS

In order to manage risks posed to pipelines, preventive and mitigation measures are considered on a case-by-case basis. These measures either reduce specific threats to pipelines or reduce the consequences of a leak or spill:

- Management of risk relies on a program that meaningfully gathers and analyzes the data related to hazards that may affect the pipeline. Based on many factors such as pipeline failure history, internal inspection of pipeline, excavation/direct assessment, susceptibility studies, models and data trending, potential risks are identified.

- The collected data is then analyzed to evaluate the potential hazards. Integrity management strategies are prepared to protect HCAs, the integrity management plan is a living process and it is allowed to evolve continuously, as a continuous improvement plan, feeding on the ever-emerging new data and performance analysis.

CHAPTER SIX

Regulatory Approach to Liquid Pipeline Risk Management

Contents

Pipeline Integrity Handbook
ISBN 978-0-12-387825-0

INITIAL APPROACH TO RISK MANAGEMENT

The pipeline industry and the Office of Pipeline Safety (OPS) have recognized the needs associated with transporting hazardous liquids through pipelines. They have addressed these risk-related concerns in a rational manner by creating a Risk Assessment Quality Team (RAQT) with OPS and API members.

RAQT was tasked to explore the potential applicability and benefits of formalized risk management programs within the liquid pipeline industry. OPS and API considered this opportunity to maximize the effectiveness of various individual efforts already initiated in the area of risk assessment and risk management and to align the goals guiding the development of risk management within government and industry.

Based on their observations, they suggested some actions that led to the API issuing guidelines for risk management of liquid pipelines. Some of the key questions that were addressed are:

1. Are the data available to support risk-based decision making?
2. How will we know if the risk-based approach is working?
3. Don't the industry and the OPS already manage risk by analyzing the incidents?
4. How would the OPS determine an operator's performance after an incident?

5. Doesn't a pipeline operator put itself in legal jeopardy by identifying very low levels of unsubstantiated risks that it doesn't immediately remedy?
6. Will pipeline companies have to display all their "warts" to the public?
7. How will the OPS be able to establish that the risk-based approach will provide equal or better protection to the public and the environment than the current regulations?
8. Is the time frame to move toward risk-based regulation of the pipeline industry too long?
9. Will the OPS now require operators to fix everything?
10. How will equity issues, such as the distribution of costs and risk, be resolved?

PARADIGM SHIFT

The responses to the above questions and suggestions resulting from these findings required some serious paradigm shifts. Some of the changes that were required of the industry can be listed as follows:

- Reactive to proactive
- Compliance-based to performance-based approach
- Prescriptive regulation to risk-based regulation
- One-size-fits-all approach to facility-specific approach
- "We are safe enough" to continuous, cost effective improvements
- Closely held information to open communication
- Fixing last event to preventing next event
- Safety versus profit to safety increases profit

- Single solution to alternatives
- Rigid rules to best-fit rules
- Exclusiveness to recognition and sharing of information as mutual need for safety was recognized.

The changes did not occur easily. A series of steps were planned to get to the point where the RAQT wanted to take the level of risk-based management of pipeline safety.

RAQT defined risk as "the possibility that an undesired event will occur." For example, a public safety risk exists if there is a chance that a nearby resident may be injured as a result of a pipeline accident.

They further used the concept of the two dimensional aspect of risk; the two aspects being frequency and consequence. They identified risk bands and assigned a frequency-of-risk versus severity-of-consequence matrix on the risk curve.

ELEMENTS OF RISK MANAGEMENT

The concept of managing the risk was discussed, and in so doing they identified the following seven elements of risk management:

1. Identify the types and sources of current and potential risks.
2. Assess the relative magnitude of the various sources of risk based on both likelihood and severity of consequences.
3. Define new practices or changes in current practice to reduce these risks.
4. Establish priorities among potential risk-reduction practices.

5. Allocate resources to select risk-reduction practices.
6. Communicate risk management decisions to key stakeholders.
7. Monitor impact of risk-reduction efforts.

The team identified that a set of analytical tools can be used to support decision-making processes. These tools can be used in a variety of ways to identify and manage risks, such as the following:

- Qualitative investigation and identification of system failure modes and mitigating actions
- Comparison and trade-offs among design, operational, inspection, maintenance, and other activities intended to reduce the frequency or consequences of pipeline accidents
- Quantitative cost-benefit analysis was possible.

Risk management programs

Not all risk management programs are the same; the simplest program feeds into the next complex program. A hierarchy of such programs was indentified in their reports:

1. Compliance-based risk management
2. Knowledge-based risk management
3. Data-based risk management
4. Model-based risk management
5. Omniscient risk management.

Each successive level of risk management supplements the information base of previous levels. This allows for a more intricate and refined information gathering system and data analysis to make the collected information more relevant.

The mandated essential requirements that pipeline operators often practice are a combination of all the five stages, but not necessarily in the same hierarchical order.

Over the years of application and data collection, the current risk-based integrity management of pipeline has evolved into a good basis for future development and possibly a safer industry.

PART TWO

Pipeline Integrity Evaluation and Engineering Assessment

CHAPTER ONE

Introduction

Contents

Pipeline Integrity Handbook
ISBN 978-0-12-387825-0

Management of pipeline integrity is the responsibility of the operator. Pipeline operators are required to maintain the integrity of their systems on a regular basis. Both liquid and gas pipelines have to meet the minimum specified regulatory requirements to keep the systems safe and in good working condition. Good maintenance policy is not only required by the mandate of regulatory authority, but is also a responsible social behavior that is expected of good corporate citizens. When we speak of a pipeline system, this includes all of the physical facilities of transportation such as the pipe, valves, appurtenances attached to the pipe, compressors for gas compression, pumps for oil, metering stations, delivery stations, launchers and receivers of pigs, and other fabricated assemblies in the transportation system.

A good pipeline integrity management system is capable of providing safe operation, accident prevention, accident control, and in case of an accident occurring, the ability to initiate quick and effective damage control and corrective measures. An effective program is capable of reducing the impact of failures to people and to the environment. Good pipeline integrity is not only necessary for an existing in-service pipeline, but must be designed into all new constructions. This involves consideration of environmental conditions in the selection of material, design of methods and facilities to reduce chances of failure and provide for measures that will allow effective control of damages, and establishment of good construction and testing practices. All these engineering efforts are built in as good integrity management. Post-construction, a well-designed program is capable of identifying and categorizing the associated risks, as well as their consequences. This may include periodic inspection, evaluation and data analysis, and initiation of preemptive corrective measures. To accomplish these goals, the operator should establish a program that must be sufficiently flexible to adapt to the changing conditions. The preemptive measures would include inspection and identification of flaws or potential causes of flaws in the pipeline system.

These flaws can be caused by both internal as well as external corrosion, mechanical damages that might include gouges, dents, grooves, and cracks, defects carried over from the fabrication, construction, or manufacturing processes.

An integrity management plan for a gas and liquid pipeline is built around information and knowledge gathered for the segment. This includes knowledge of technical and surrounding environment. The internal inspection and testing data, as well as the information collected about outside conditions of the pipeline, from excavations performed on the pipeline system or other tools used for this purpose.

This detailed collection of data about the pipeline segment enables the integrity management team to monitor and conduct a full assessment. Using this database, the integrity management plan is designed and implemented to assess and address the associated risk. This proactive engineering effort allows the identification of risks associated with each potential defect that exceeds determined limits of acceptance. This level of engineering involvement allows for timely identification and repair of those defects in the pipeline.

The focus of any integrity management plan should be to identify and correct low-level pipeline damage and deterioration before major threats arise demanding extensive repairs. The objective should be to avoid accidents along the pipeline, but extra steps are necessary in reducing the risk level and preventing failures in high consequence areas (HCAs).

The integrity management plan should include risk assessments to comprehensively evaluate the range of threats to the pipeline segment and their impact on adjoining HCAs. Identifying potential threats by type, which over time may further deteriorate the integrity of the pipeline and become major hazards, allows for preemptive action to be taken. These threats generally fall into one or more of the following categories:

- Metal loss or corrosion
- Pipe deformation, such as denting caused by third-party activities in the vicinity of the pipeline
- Failure of material due to the manufacturing or forming processes
- Failure related to exposure to natural environments
- Failure due to incorrect operating conditions.

As discussed earlier in Part 2 of this book, the risk analysis involves the use of data within an integrity assessment program. This data is gathered from various sources such as those listed below. However, the collection of data is not limited to these sources alone:

1. Original construction records
2. Pipeline alignment sheet records
3. Personnel interviews
4. Geological survey records such as Quadrangle USGS maps in the USA
5. Digital elevation models
6. Historical data
7. Leak and incident data/reports
8. Operating characteristics
9. Corrosion monitoring
10. Cathodic protection surveys
11. Transported product information
12. Digital maps delineating HCAs.

PIPELINE INTEGRITY MANAGEMENT

The objective of pipeline integrity management is to improve pipeline safety through:

- Assessing risk associated with the system
- Identifying HCAs (a summary of HCA assessment for gas pipelines is given in Table 2-1-1 below)
- Improved integrity management program within company
- Improving the regulatory authorities' review for the adequacy of integrity programs and plans
- Providing assurance to the general public about the safety of pipelines.

Table 2-1-1 Summary of HCA Assessment Requirements 49 CFR 192.

	In the following situations, assessment of pipeline HCA area is required					
	Operating $\geq$ 50% SMYS		**Operating $\geq$ 30% SMYS**		**Operating $<$ 30% SMYS**	
Baseline assessment method	**Maximum re-assessment interval**	**Assessment method**	**Maximum re-assessment interval**	**Assessment method**	**Maximum re-assessment interval**	**Assessment method**
Pressure testing	7	CDA	7	CDA	Ongoing	Prevention & mitigation (P&M) measures
	10	Pressure test or ILI or DA		Pressure test or ILI or DA	20 years	Pressure test or ILI or DA
		Repeat inspection cycle every 10 years		Repeat inspection cycle every 15 years		Repeat inspection cycle every 20 years
In-line inspection (ILI)	7	CDA	7	CDA	Ongoing	Prevention & mitigation (P&M) measures
	10	ILI or DA or pressure test				
		Repeat inspection cycle every 10 years	15	ILI or DA or pressure test		
				Repeat inspection cycle every 15 years	20	ILI or DA or pressure test

As we can deduce from the above, the principal objective of integrity management is the assessment of pipelines in HCAs. Based on identification of HCAs, establishing a good integrity management system will ensure a company earns the public's confidence in the safety of the pipeline running in their neighborhood. It should also be able to establish frequent and meaningful communication channels between the operators and the regulatory authorities. The aim of this is to instill a sense of confidence in the general public that, in the case of any accident occurring, necessary steps will be initiated within a reasonable time and protection will be available for the HCA.

HIGH CONSEQUENCE AREA ANALYSIS

As is stated above and as is now common knowledge, the regulations for pipeline operators within the oil and gas pipeline industry are becoming increasingly rigorous, especially in the fields of pipeline integrity and emergency response protocol. The reporting requirements about HCAs have opened up the need for better tools that would aid in the management and assessment of both directly and indirectly affected HCAs. The aim of HCA analysis (Figure 2-1-1) is to identify all segments of a pipeline system having the potential to affect an HCA, either directly or indirectly.

Direct and indirect HCAs

The key difference between direct and indirect HCAs is as follows:

- Directly affected HCAs are those that sit on and intersect the pipeline centerline.

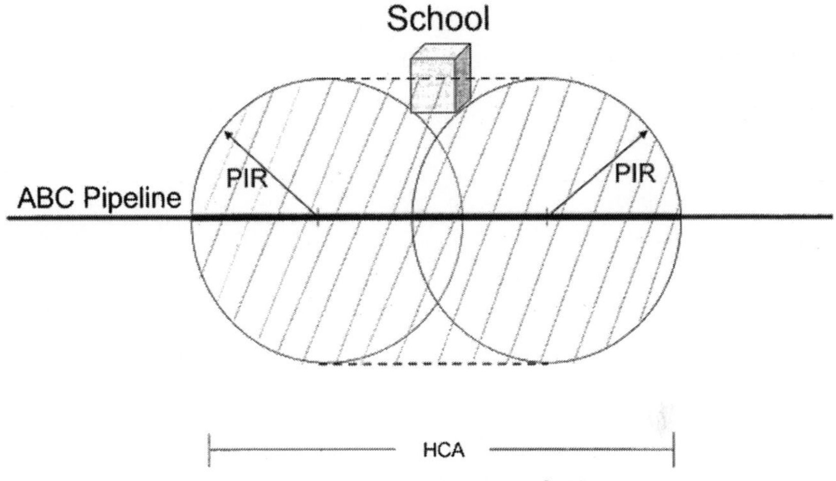

Figure 2-1-1 Determination of HCAs.

- Indirectly affected HCAs are those that are affected by liquid pooling or thermal radiation or that fall within a predefined analysis area (i.e., a risk-based distance).

Determining directly affected HCAs is a relatively straightforward analysis and one that the pipeline industry can easily determine.

However, effective determination of indirectly affected HCAs within an analysis area is not so easy to accomplish. The correctness of the analysis and determination is often a challenge to the pipeline industry. There are a range of HCA analysis tools available to improve this process, and they provide a fair degree of accuracy.

HCA analysis automation

HCA analysis tools are available from technology enhancing companies to support the assessment and reporting of both directly and indirectly affected HCAs within simple or complex analysis areas.

These tools allow the user to define an analysis area by utilizing a selected offset distance field or a complicated predefined polygonal geometry (such as a pre-determined spill area) that cannot be represented as a single offset value. Some of these tools are user friendly and can be easily maneuvered with the use of point, line, and polygon features as defined HCAs and the output reports direct HCAs and indirectly affected HCAs.

Identification of HCAs

The pipeline operators have the responsibility to identify the HCA in relation to their pipelines or segments of pipeline. They should develop programs to classify HCAs. They should develop a plan of inspection and monitor the pipeline segments to identify and protect failures.

Developing a framework that identifies how each element of the integrity management program will be implemented is an essential step. The integrity management program must include the following elements:

- A process for determining which pipeline segments could affect an HCA
- A baseline assessment plan

- A process for continual integrity assessment and evaluation
- An analytical process that integrates all available information about pipeline integrity and the consequences of a failure
- Repair criteria to address issues identified by the integrity assessment method and data analysis
- A process to identify and evaluate preventive and mitigation measures to protect HCAs
- Methods to measure the integrity management program's effectiveness
- A process for review of integrity assessment results and data analysis by a qualified individual.

Each of these areas must be addressed in the developed specification. The operator must perform periodic integrity assessments, as part of a continual integrity evaluation and assessment strategy. Such assessment should not exceed 5-year intervals. The regulatory requirements state that certain defects identified through internal inspection must be repaired within defined time limits. These are outlined in the following sections.

Immediate repair

The following conditions require immediate repair:

1. Detection of metal loss greater than 80 percent of nominal wall thickness
2. Calculated burst pressure less than maximum operating pressure at anomaly
3. Dent on pipe topside that shows metal loss or cracking or that becomes a stress riser
4. Any other anomaly that demands immediate attention.

Repair within 60 days

Other severe anomalies that may be attended to and rectified within 60 days are:

1. Dents in the top section of pipe greater than 3 percent of the nominal pipe diameter. For pipes less than 12 inches in diameter, this limit is 0.25 inches
2. Bottom dents with any indication of metal loss, cracking, or stress riser.

Repair within 180 days

Less severe anomalies that may be repaired within 180 days include:

1. Dents greater than 2 percent of the nominal pipe diameter located in the proximity of pipeline girth weld or longitudinal seam
2. Dents that are more than 0.25 inches in a 12-inch diameter pipe and located in the proximity of pipeline girth weld or longitudinal seam
3. A dent on the top of the pipe that exceeds 2 percent of the pipe diameter
4. A dent that is more that 0.25 inches in pipes of up to 12 inches in diameter
5. A dent at the bottom segment of the pipe that exceeds 6 percent of pipe diameter
6. An area of general corrosion with predicted metal loss that exceeds 50 percent of nominal wall thickness
7. Predicted metal loss exceeding 50 percent of nominal wall thickness, at a crossing of another pipe, or where there is

an area of widespread circumferential corrosion, or in an area that could affect a girth weld

8. Calculated operating pressure falls below the maximum operating pressure at anomaly
9. A gouge or groove that exceeds 12.5 percent of nominal wall thickness.
10. Corrosion of, or along, a longitudinal seam weld
11. A potential crack or an excavation determines that there is a crack.

The above discussion points to the fact that defects are identified and measured correctly and grouped in specific categories to apply the regulatory mandates. Figures 2-1-2 to 2-1-6 depict some of the defects and their measurement in

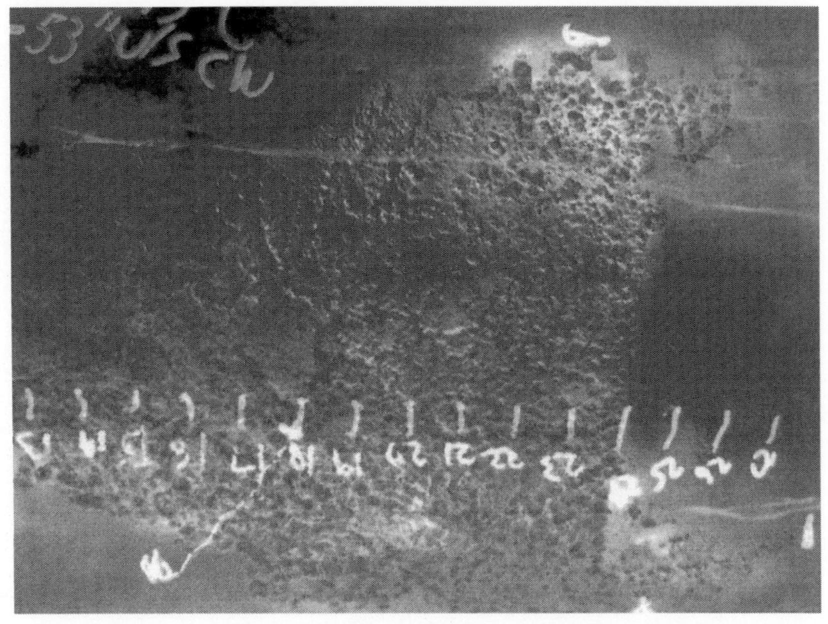

Figure 2-1-2 External corrosion and measured pit depths.

Figure 2-1-3 Measuring the depth of a dent.

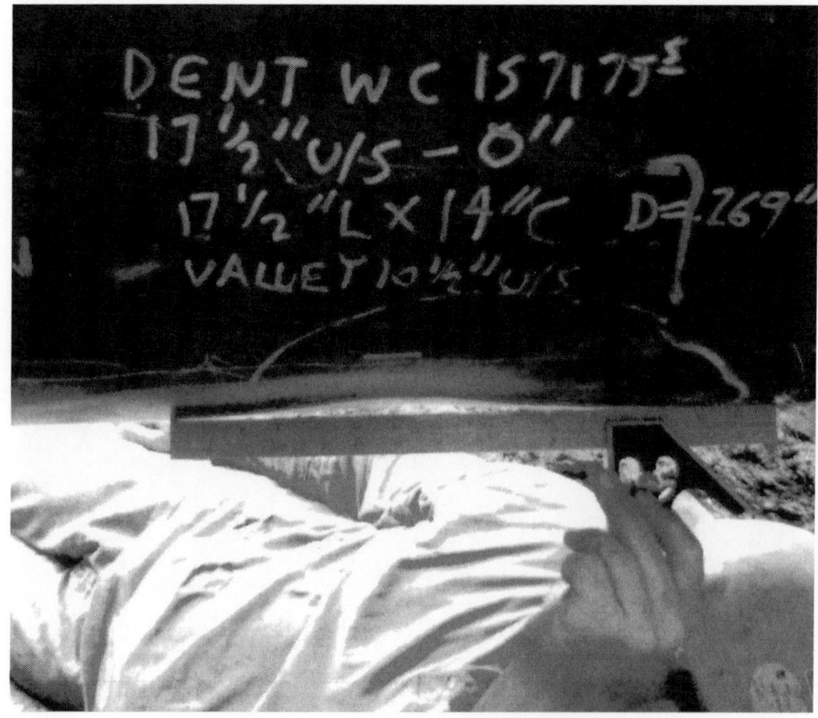

Figure 2-1-4 Measuring the length of a dent.

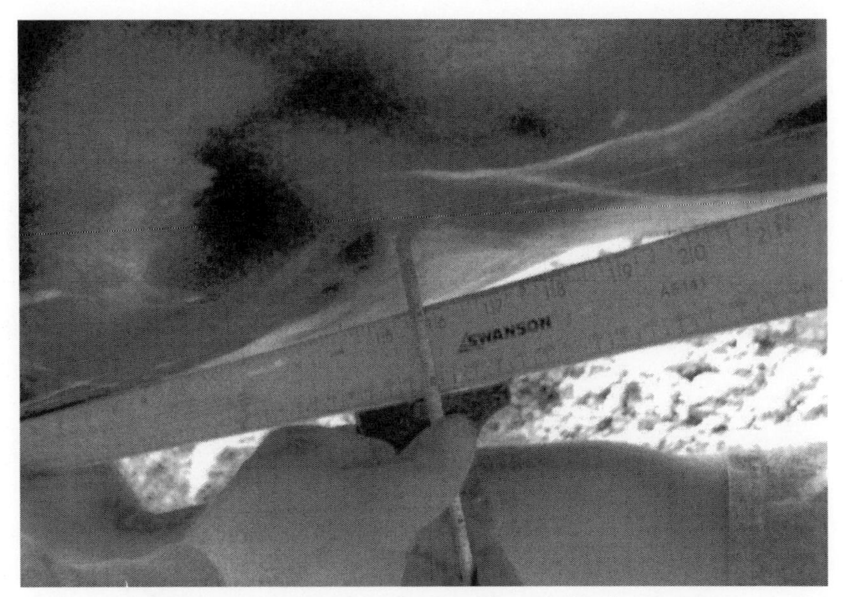

Figure 2-1-5 Measuring the length and depth of a pit.

Figure 2-1-6 Measuring the length and depth of a dent near the girth weld.

the field. A good integrity management program will integrate all available information, including the following minimum data:

- The potential for mechanical damage by excavation or any other outside force damage, taking into consideration potential new developments along the pipeline
- An evaluation of the impact of potential release on the HCA, e.g., drinking water intake, damage to sensitive ecosystem
- Data collected from cathodic protection surveys
- ROW patrolling once a year for class 1 and 2 locations, every 6 months for class 3 locations, every 3 months for class 4 locations, and other maintenance and surveillance activities
- Other regulatory bodies' mandated data collected relating to the oil and gas pipeline integrity.

The above discussion brings out the importance of understanding pipeline defects, their identification, and their potential impact on pipeline integrity. In the subsequent chapters of this section we will try to address these issues.

LIQUID SYSTEMS HIGH CONSEQUENCE AREAS

The determination of HCA for liquid systems is based on the same principles as for gas pipelines, except that due to the inherent nature of liquid and its potential to spread in the environment, some differences occur. The sensitive areas are defined as sources of drinking water or ecological resource areas that are especially responsive to environmental damage from a hazardous liquid pipeline release. The US federal government has applied this definition to identify HCAs and has made maps depicting these locations for pipeline operators. Operators are also responsible for independently evaluating information about the areas around their pipelines to determine whether a pipeline accident could affect a nearby, but not adjacent, HCA.

Integrity management approach to liquid pipelines

The integrity management plan for liquid pipes is based on the information and data collected for the pipeline or its segment. As with gas pipelines, this includes technical knowledge and information about the surrounding environment. It also includes inside-the-pipeline inspection and testing data, as well as information gathered from outside the pipeline from cathodic protection data and from excavations performed on the pipeline system.

The detailed knowledge obtained from the various sources of collected data on a pipeline segment enables proper monitoring and assessment of any defects. The integrity management plan is designed and implemented to assess and address, in a proactive manner, the risk associated with each potential defect that exceeds acceptable tolerances.

As for the gas transportation pipelines, the approach for liquid pipelines also involves identification and timely repair of defects in the pipeline. The use of engineering assessments and regulatory specifications to determine the acceptance level is common practice.

The objective of any integrity management plan should be to identify and correct pipeline defects that would have a low impact on the environment before they become a major repair situation. Risk assessment is a major part of any integrity management plan; this assessment should comprehensively evaluate the range of potential threats to pipeline segments and consequences to any nearby HCAs. The types of potential threats include hazards or damage that, over time, deteriorate the pipeline. The risk analysis involves the use of data collected in the same manner as for gas transportation pipelines, discussed above.

CHAPTER TWO

Pipeline Defects and Corrective Actions

Contents

Pipeline Integrity Handbook
ISBN 978-0-12-387825-0

One important part of pipeline integrity management activity encompasses the repair and maintenance of anomalies by the maintenance crew as they carry out their periodic scheduled maintenance activity. The identification and inspection of these anomalies is an important component of the above activity.

The procedures are established within the integrity management plan to investigate and analyze all failures and accidents. However, it is the keen eyes of the operators and maintenance crew that are the first line of preventive defense from a failure. This aspect of activity emphasizes the need for operator qualification (OQ) for specific functions of crew members.

In addition to the advanced inspection tools, knowledge of pipeline defects and how to conduct both immediate and scheduled repairs assumes critical importance. The training and education of operators play critical roles in the effectiveness of their ability to identify a potential defect and to prevent a potential failure, while patrolling the pipeline, inspecting the crossings, surveying a leakage, and identifying potential fire hazards. The knowledge and ability to control other activities that may have a potentially damaging effect on the pipeline such as construction activity in the proximity of pipeline, a farmer plowing his farm above the buried pipeline, or blasting activity, the impact of which may adversely affect the stability of soil or even cause direct damage to the pipeline. It is very critical for the safety and integrity of oil and gas pipelines that we are aware of these defects, whether they are identified by visual inspection or reported by any of the various inspection means discussed in the previous chapter.

THE MOST COMMON DEFECTS

The most common defects are those caused by external damage. They can be one of the following types: dents; grooves; gouges; and arc strikes (also referred to as arc burns).

Dents and grooves on a pipeline are injurious if their depth is more than 10 percent of the nominal wall thickness. This is, however, subject to the pipe wall calculation of the design pressure. Dents are an abrupt change in the profile of the pipe surface. The smooth profile of dents does not require repair. However, if their depth is greater than 6 percent of the nominal pipe wall thickness, or if dents contain any stress concentrator such as scratches, grooves, gouges, or arc burns, or if dents occur on the longitudinal welds, they are to be evaluated and repaired. Typical preventive and mitigative action recommended for gas pipelines is given in Table 2-2-2 below.

ASSESSING REMAINING STRENGTH OF CORRODED PIPE

The pipeline industry had used the ASME B31G criteria to evaluate corroded pipe for removal or repair or for leaving it in service if the metal loss was within safe limits of calculations.

The B31G criteria helped pipeline operators to avoid many unnecessary cutouts. However, the conservative approach of the B31G criteria did account for some cutouts that were not necessary and would not have compromised the safety of the pipeline. Because of this, there was a need for the establishment of new and improved criteria to make the remaining strength more efficient. The typical data used for calculation of strength of pipeline according to B31G criteria is given in Table 2-2-1.

Determining the remaining strength of corroded pipelines based on ASME B31G

MODIFIED CRITERIA FOR EVALUATING THE REMAINING STRENGTH OF CORRODED PIPE

The modified criteria were established with the objective to reduce excess conservatism without creating an unsafe condition. The initial 47-burst test results were revisited and the 31G criteria were validated, with the modified criteria being assessed against those test conditions. This was to establish that the modified criteria would provide adequate means to predicting the effects of metal loss on the remaining strength of the corroded pipe. In addition, through efforts of individual companies, 39 additional burst test results were used for the validation effort. The expanded database established that the modified criteria had an adequate margin of safety. The modified criteria could be used with detailed measurements of the metal loss and successive trial calculations to predict a

Table 2-2-1 Typical Data required for calculation

Nominal outside diameter of the pipe (D):	in
Nominal Pipe Wall Thickness (t):	in
Measured longitudinal extent of the corroded area (LM):	in
Measured maximum depth of the corroded area (d):	in
Maximum Allowable Pressure (MAOP):	psi
Specified minimum yield Strength (S):	psi
Appropriate design Factor from ASME B31.4, ASME B31.8, or ASME B31.11 (F):	0.72
Temperature derating factor:	1
Output information	
Variables	**Values obtained**
Percentage Pit depth	
Max allowable Longitudinal (L) corroded area, colinear with LM in inches	
Design Pressure in psi	
Safe maximum pressure for corrode area (p') psi	

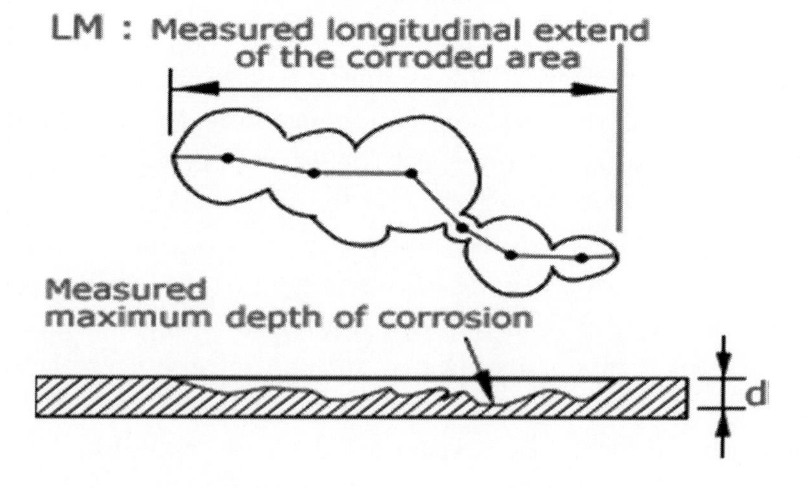

Table 2-2-2 Preventive and Mitigative Measures Addressing Time Dependent and Independent Threats to Transmission Pipelines Operating Below 30 Percent SMYS, in HCAs.

Threats	CFR 49 192 requirements		Additional preventive and mitigative measures
	Primary	**Secondary**	
External corrosion	455 (Gen. post-1971)		**For cathodically protected (CP) pipeline:** Conduct an electrical survey using an indirect examination tool or method at a minimum of 7-year intervals.
	455 (Gen. pre-1971)		The results are to be utilized as part of an overall evaluation of CP system and corrosion threats for the covered segments.
	459 (Examination)		
	461 (Ext. coating)	603 (Gen. oper.)	
	463 (CP)	613 (Surveillance)	**For CP pipeline:** Conduct an electrical survey using an indirect examination tool or method at a minimum of 7-year intervals.
	465 (Monitoring)		
	467 (Electrical isolation)		
	469 (Test stations)		
	471 (Test leads)		
	473 (Interference)		
	481 (Atmospheric)		
	485 (Remedial)		
	705 (Petrol.)		

(Continued)

Table 2-2-2 Preventive and Mitigative Measures Addressing Time Dependent and Independent Threats to Transmission Pipelines Operating Below 30 Percent SMYS, in HCAs.—cont'd

Threats	CFR 49 192 requirements		Additional preventive and mitigative measures
	Primary	Secondary	
	706 (Leak survey) 711 (Repair gen.) 717 (Repair perm.)		
Internal corrosion			
Third party damage	7	CDA	

Figure 2-2-1 Measurement of corroded area.

minimum failure pressure for an area of metal loss based upon its "effective" area. Used in this mode, the modified criteria further reduced the excess conservatism associated in the 31G criteria. This modified criteria analysis approach was developed into a PC-based program called RSTRENG®.

Pipeline operators are required under 49 CFR192 and 49 CFR195 to use either the RSTRENG® or the ASME/ANSI

B31G criteria to evaluate the remaining strength of corroded pipe. The RSTRENG® criteria are less conservative than the B31G criteria, and therefore have the potential to reduce unnecessary pipe cutouts. RSTRENG® permits a determination of the metal-loss anomalies that may safely remain in service at the current maximum operating pressure. For anomalies that exceed the recommended allowable size, the modified criteria will establish the appropriate pressure reduction to maintain an adequate margin of safety for all cases in which the reduced pressure level exceeds 55 percent of specified minimum yield strength (SMYS).

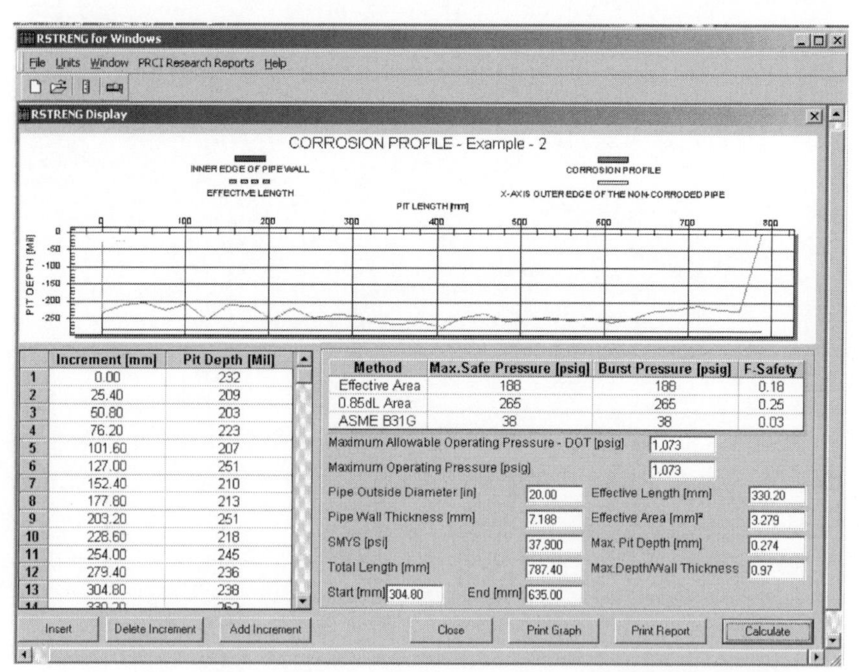

Figure 2-2-2 Typical Microsoft Windows screen of RSTRENG® program.

PRCI continues to validate the modified RSTRENG® criteria. The latest results of 129 new tests involving corroded pipe or pipe samples containing corrosion-simulating defects provide both qualitative and quantitative validation of the RSTRENG® technology.

The subsequent improvement in the RSTRENG® program has made it more user-friendly in Windows format (Figure 2-2-2) and includes:

- Internationalized format
- All supporting technical documentation
- Results that are compared with the B31G criteria.

Pipeline Reliability Assessment

Contents

Pipeline Integrity Handbook
ISBN 978-0-12-387825-0

Another aspect of pipeline integrity management is the assessment of system reliability over the age of the pipeline. In order to assess the aging effects through the pipe's lifetime, a reliability assessment is carried out for the pipeline or its segment. This two-step method determines the influence of residual stresses in uncorroded pipeline:

1. This step establishes the sensitivity of system parameters.
2. In the second step, the residual stress model is coupled with the corrosion model in order to assess the aging effects through the pipe's lifetime.

We discuss these two steps in subsequent paragraphs.

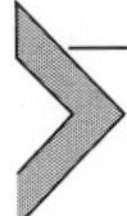

INFLUENCE OF RESIDUAL STRESS ON PIPELINE RELIABILITY

Operating mean pressure

The assessment of the residual stress effect is carried out by evaluating the reliability of new uncorroded pipelines, which are assumed to be free from any flaw. In such hypothetical conditions, the residual stress is linear according to Crampton's law, in an inert environment:

$$\sigma_{resc}(r_t) = -70\left\{1 - (2r_t/t)\right\}$$

where r_t is the radial coordinate of the considered point in the wall thickness (i.e., $0 < r_t < t$), measured from the outer radius.

The linear fitting allows us to find an equilibrium state of compressive and tensile residual stresses. The reliability is evaluated as a function of 20 independent random variables involving materials, geometry, the coefficients, loadings, and residual stress.

Use of software to determine reliability allows computing the reliability index β and the failure probability P_f, when internal pressure is up-rated. It may be noted that, often, the decrease in reliability due to residual stress at low pressures is more noticeable than on increased pressure.

VARIABLE SENSITIVITIES

The sensitivities α^2 or the relative contribution of the random variables to the pipe reliability for gas pressure increases. The most important variables for pipeline safety are the thickness, the yield stress, and the applied pressure. When the mean value of residual stress is 2175 psi (15 MPa) and the pressure is up-rated, the thickness and the yield strength contribute to about 50%; the remaining part is equally divided among the other parameters. However, for a system operating at low mean pressure, all parameters contribute significantly.

The factors α^2 can be plotted for the full range of operating gas pressures, in order to show the evolution of variable importance. Except for the internal pressure, for which steeply increasing sensitivity expressed by α^2 is observed, this signifies that the sensitivities of all the other variables have low variations. In particular, a comparison of the sensitivities of the residual stress and the internal diameter shows opposite evolutions. At low operating pressures, these sensitivities are almost constant. Above the mean pressure of 2900 psi (20 MPa), the sensitivity of the internal radius increases, while the sensitivity of residual stress decreases. When the operating pressure is very high, the pressure sensitivity becomes dominant and therefore squeezes out most of the other variables (except the thickness and the yield strength). Thus, the five parametric variables that emerge are:

- Yield strength
- Wall thickness
- Internal pressure
- Internal radius
- Residual stress.

GAS PRESSURE FLUCTUATION

The reliability index β is a function of the coefficient of variation of pressure ρ_P for operating mean pressure.

It is observed that β decreases quasi-linearly in cases with and without residual stress. When residual stress is not considered, for a mean pressure of 4351 psi (30 MPa), β decreases, as the coefficient of variation increases from $\rho_P = 0.1$ to 0.3. Similar observations are made when residual stress is taken into account; in this case, the target reliability for $\rho_m = 30$ MPa corresponds to $\rho_P = 0.1$.

In all cases, for a given failure probability, if the dispersion increases, the operating pressure should be reduced. Moreover, the slope of the curves obtained with residual stress considerations is less than for the case without residual stress. This indicates that residual stresses could become "beneficial" as they reduce the influence of pressure fluctuations. For high pressures, the failure probability is still high with increasing ρ_P, as β falls well below the target. Even with severe control of pressure fluctuations, it is not possible to maintain safety at acceptable levels.

RESIDUAL STRESS DATA

In this section, the influence of residual stress parameters, mean, and coefficient of variation are considered in the reliability assessment. The evolution of the reliability index is a function of the mean value of outer radius residual stress. For simplicity, it will be referred to here as mean residual stress. Increasing the mean residual stress implies a linear drop of the index β. The slope is nearly the same for different operating pressures. When the mean residual stress varies from 0 to 15,954 psi (0 to 110 MPa), the index β drops by nearly two graduations. The coefficient of variation of residual stress $\rho_{\sigma res}$ shows much less sensitivity in the pipe reliability. In fact, as the pressure increases, the residual stress uncertainty loses its importance. A practical conclusion is that a reasonable accuracy is deemed sufficient for characterization of the residual stress dispersion.

CORROSION EFFECTS

The reliability assessment of underground pipelines with active corrosion is established with the model given as follows:

$$t_c = kT^n$$

where t_c is the thickness of the corroded layer, T is the elapsed time, and k and n are the corrosion constants.

For atmospheric pressure with parameters $k = 0.066$ and $n = 0.53$, the corroded layer in a lifetime of 50 years is equal to 0.52 mm. In more aggressive environments, the corrosion process is activated by increasing the parameter k.

The aim should be to determine the pipeline's reliability with regard to corrosion rates together with residual stress effects. As the corrosion process is a time-variant process, the use of random variable theory is only possible under the hypothesis of monotonically increasing failure probability. This hypothesis would allow evaluating the time-dependent probability by simple computation of the instantaneous failure probability. From a physical point of view, this is compatible with the corrosion process, but implies that no significantly large jump of loading can occur during the pipe's lifetime. Moreover, the computation of instantaneous failure probability with extreme loading values gives an upper bound to the time-variant failure probability.

As expected, the reliability index β decreases with pipe aging, and correspondingly, the failure probability increases with time T.

High corrosion rates produce a very large decrease in pipeline safety, especially in the early stage of its lifetime. The nonlinearity of these curves is more significant for higher corrosion rates compared to atmospheric corrosion. During the reference lifetime of 50 years, the influence of residual stresses decreases with time, since after 20 years with high corrosion, the reliability level becomes less sensitive and converges to that obtained for the case without residual stress. The explanation for this is that when corroded layers are consumed, the pipe wall thickness reduces and the residual stresses are relaxed, and consequently their effect is reduced.

One of the major pieces of information from these models is to supply the design engineer and operators with a realistic image of the pipeline risk of failure at various lifetime instances with regard to corrosion rates. As an example, for a moderate corrosion rate, for a 30 MPa pressurized pipe, the probability of failure can be acceptable for an operation time of 50 years, whereas for a 5801 psi (40 MPa) pressurized pipe this is not true, since after 20 years, the corroded pipeline presents a very large risk of failure.

PART THREE

Pipeline Material

CHAPTER ONE

Introduction

Pipeline Integrity Handbook
ISBN 978-0-12-387825-0

In a typical pipeline project a variety of materials are used, including line pipe, valves, induction bends, fittings, and various forgings such as flanges, weldolets, anchor forgings, and gaskets. These materials are subject to challenges from environmental and mechanical conditions imposed on them by the design and service requirements. These conditions develop stresses that limit materials' performance. In this part of the book we will discuss the materials commonly used in pipeline projects, their inherent properties, and how they are affected by the environment in which they are to perform. While on the subject of materials selection and their performance, the discussion will also suggest some of the protective measures that can prevent premature failures and point out the limits of these protective measures.

The process of selecting material for a pipeline system has two clear paths; they are not mutually exclusive paths but are considered rather simultaneously; often an engineering compromise is made in balancing the two objectives:

1. The type and shape of the parts
2. The physical and mechanical properties of the material these parts are made from.

In the larger scale of the picture, a pipeline system will have the pipe, valves, pumps for liquid transportation and compressors for gas transportation, flanges, fittings, and fabricated equipment like vessels, tanks, boilers, heat exchangers, and many more. Some of these parts may be made of fiberglass-reinforced plastics, plastics, and sometimes non-ferrous materials, but predominantly various grades of steel are used.

These general material types may have different compositions and may be produced by different production processes. The steel, which constitutes the bulk of material used in pipeline systems, is available in various grades and types. The selection is from various grades of steel that are produced in different ways and the production process of each grade or group of material varies significantly. For example, pipe in most cases is produced by rolled plates that are welded, or may be drawn and extruded out of steel stock to produce seamless pipes.

A valve's body may be made of either cast-steel or forged steel. Flanges, with rare exceptions, are almost always made of forged steel.

Fittings are often stamped out of wrought steel, or in some cases forged steel. They may be of seamless or welded construction.

Pump casings are usually made of cast steel or may, in some cases, be made of forged steel. However, they are (like valves) an assembly of various grades of steel and alloy steels produced from cast, forged, and wrought steel material, which are further processed by machining and other secondary processes.

In all of the above cases, the primary processes may be further supported by one or a combination of secondary processes including rolling, stamping, machining, and welding. The ability to understand the sequence of these processes and a knowledge of the secondary processes helps understand the material in more detail. The manufacturing of specific products entails a specific sequence of operations; knowledge of these can help develop full understanding of the manufacturing process and the material itself.

The determining factors in the selection of materials for pipeline systems are the environment where the candidate material will be put into service, its design life, and the cost.

In the first category, the environment of material performance, the factors to consider are the pressure and temperature of the system, the corrosiveness of the fluid carried by, or exposed to, the pipeline and components. The exposure implies the conditions that the system will encounter in order to enable it to function for optimum results.

The mechanical properties of materials would consist of tensile strength, yield strength, ductility, and toughness as determined by impact tests and supporting tests like hardness and micro-structure evaluation. In cases where the material is subject to cyclic stress, additional toughness tests and determination of K_{IC}, critical crack tip opening displacement (CTOD), and critical J values of material and weld may also be specified.

In the following chapters we will discuss various pipeline materials.

CHAPTER TWO

Line Pipes

Contents

Pipeline Integrity Handbook
ISBN 978-0-12-387825-0

The manufacture, testing, and classification of line pipes are predominantly controlled by API 5L, a specification which is now also an ISO specification (ISO 3183). It is strongly advised that the pipeline engineering team obtain the latest edition of this specification and use it for design, construction, procurement, and quality control of their projects. The API pipes for pipeline systems are primarily classified on the basis of the material's yield strength and the process of making the pipe. In general, there are two processes: seamless pipes and welded pipes. The welded pipes are further classified on the basis of the welding process. The other classification for all pipes is based on their physical attributes, such as diameter and wall thickness.

Identification is on the basis of an alphanumeric system (e.g., X42, X56, X70, etc.). Some grades, such as Grade A and Grade B, are the exceptions, as they do not reference the yield strength but, like all other grades, they too have specified minimum yield strength (SMYS). Grade A steel has SMYS of 30,000 psi and Grade B has SMYS of 35,000 psi.

They are further classified on the basis of product specification level (PSL), and there are two levels of product quality, PSL 1 and PSL 2. Pipes produced to meet PSL 2 requirements are different from PSL 1 pipes on the basis of the mandatory impact test. Imposition of upper limits to tensile and yield strength and some welding processes, specifically a PSL 2 pipe, may not be low frequency electric resistance welded (LFW) but must be welded by a high frequency welding (HFW) process, which is different from electric resistance welding

(ERW). This standard term is reclassified in API 5L as EW. The text below explains the difference between LFW and HFW electric welded pipes.

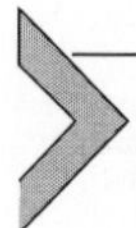

ERW PIPES: LOW FREQUENCY VERSUS HIGH FREQUENCY WELDING

ERW has, for many years, been used for making longitudinal seam welds in steel line pipe, principally for use in low-grade pipeline applications. The advent of high frequency induction (HFI) techniques has led to significant improvements in weld quality, which in combination with greater control of the raw material chemical composition, has led to the production of high quality line pipe suitable for more stringent applications in oil and gas pipelines.

Although this product form has been used for many years in sour service environments, and in the North Sea for sweet service applications, there is some resistance to the use of ERW line pipe, particularly in stringent service applications. This lack of confidence is generally based on historical problems related to the reliability of ERW, the pressure reversals and preferential weld line corrosion, and susceptibility to stress corrosion cracking.

The quality that can be achieved with modern HFW line pipe has improved dramatically, with improvements in strip quality and a greater understanding of the welding process and non-destructive testing technology. It offers closer dimensional control, which is of great value in pipe laying, and gives potentially significant cost savings over seamless pipe for similar applications.

The advantage of HFW over low frequency ERW processes is that the HFW has a very narrow heat affected zone (HAZ), so the current flow and thus the heat is limited to a very small area. In this process, two proximity conductors under a magnetic core are placed opposite each other on the edges of the pipes to be welded (Figure 3-2-1). This proximity causes the edges to heat and, as they are heated, a force is applied to bring the two faces together and a weld is made. To heat a similar area with conventional ERW (EW in API 5L) would require a significantly higher current, and would result in a wider HAZ.

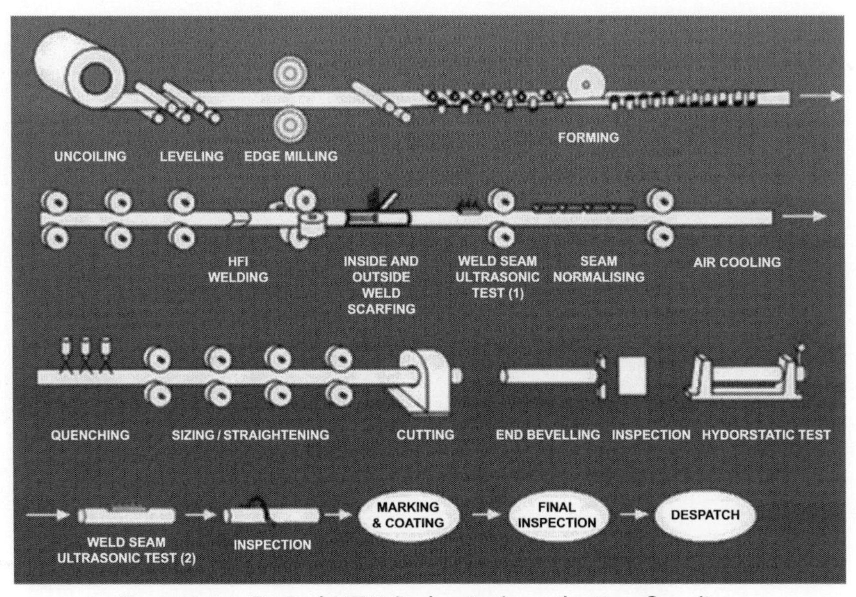

Figure 3-2-1 Typical HFW (induction) production flow line.

SUBMERGED ARC WELDED LINE PIPES

Other processes may include submerged arc welding (SAW); with this process the welds are made from both inside as well as outside. Welding from both sides generates another term, double submerged arc welding (DSAW), often used in the industry. The weld positioning is also a factor in specifying the type of pipe; for example, a straight long weld along the length of the pipe is called longitudinal submerged arc welding (LSAW). If the pipe is made by helically twisting a long steel coil, the weld will be along the abutting faces of the coil forming the pipe. Such welding is conducted by SAW processes from both inside as well as outside the seam and this is called helically welded pipe (HSAW). Both these terms have been rechristened in the new edition of API 5L as SAWL and SAWH; note the relocation of the adjectives "helical" and "longitudinal" in the new names. Table 3-2-1 gives various processes used to produce line pipes.

CLASSIFICATION OF LINE PIPES

Line pipes are also classified on the basis of their mechanical properties. This classification is based on the concept of SMYS, which is often the basis for all engineering design calculations. It may be noted that the SMYS of a grade is a fixed number and it is used to define pipe as X42, X56, or X70, etc.

Table 3-2-1 Acceptable Line Pipe Manufacturing Processes.

Type of pipe	Description or definition
SMLS	Seamless pipe produced by hot forming process; after hot forming, cold sizing and finishing is carried out
CW	One longitudinal weld produced by continuous welding
LFW	Pipe produced with low frequency (<70 kHz) EW process
HFW	Pipe produced with high frequency (>70 kHz) welding process
SAWL	Submerged arc welding process – longitudinal seam
SAWH	Submerged arc welding process – helical seam
COWL	Longitudinal seam pipes produced by a combination of metal arc and submerged arc welding processes
COWH	Horizontal seam pipes produced by a combination of metal arc and submerged arc welding processes
Double-seam SAWL	Pipe made in two halves and both longitudinal welds are made by SAW process
Double-seam COWL	Pipe made in two halves and both longitudinal welds are made by combined metal arc and SAW processes

Table 3-2-2 lists the details of various grades of steel according to API 5L/ISO3183 and the required minimum tensile and yield strength of materials and welds.

The API 5L gives tolerances on pipe dimensions, such as diameter, out of roundness, and wall thickness in tables of various editions. In the 44th edition these tables include Table 9 (Permissible outside diameter and wall thicknesses), Table 10 (Tolerances for diameter and out of roundness), Table 11 (Tolerances for wall thickness), and Table 12 (Tolerances for random length pipe, etc.). It may be noted that these

Table 3-2-2 Properties of Steel Line Pipe.

Pipe grades	Pipe body seamless or welded pipes				SAW, EW, & COW weld
	Minimum yield strength MPa (psi)		Minimum tensile strength MPa (psi)		Minimum tensile strength MPa (psi)
A25 (L175)	175	25,400	310	45,000	310 (45,000)
A25P (L175P)	175	25,400	310	45,000	310 (45,000)
A (L210)	210	30,500	335	48,600	335 (48,600)
Grades above are available in PSL 1 only					
Grades below are available in both PSL 1 and PSL 2					
B (L 245)	245	35,500	415	60,200	415 (60,200)
X42 (L290)	290	42,100	415	60,200	415 (60,200)
X46 (L320)	320	46,400	435	63,100	435 (63,100)
X52 (L360)	360	52,200	460	66,700	460 (66,700)
X56 (L390)	390	56,600	490	71,100	490 (71,100)
X60 (L415)	415	60,200	520	75,400	520 (75,400)
X65 (L450)	450	65,300	535	77,600	535 (77,600)
X70 (L485)	485	70,300	570	82,700	570 (82,700)

tolerances are generalized acceptable levels or ranges; in some specific cases, the engineering team must consider either accepting the given tolerances or changing them to suit their specific requirements. This includes changes in chemical composition of various elements as well as variations in mechanical properties and dimensional tolerances.

PSL 1 VERSUS PSL 2

One of the main differences between PSL 2 and PSL 1 pipes is the mandatory Charpy V notch (CVN) absorbed energy requirements for the pipe body of PSL 2 pipes. Table 8 of API 5L gives the minimum requirements for various grades of PSL 2 pipes from 20-inch (508 mm) diameter to 84-inch diameter pipes. These values are for normal service line pipes. For more stringent requirements, like cold temperature service, and where ductile fracture conditions are expected, higher values may be calculated and specified.

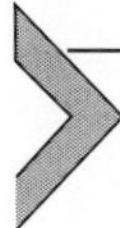

DETERMINATION OF PERCENTAGE SHEAR THROUGH DWTT FOR PSL 2 WELDED PIPES

The pipe material is expected to be sufficiently ductile to resist brittle fracture and establish that the material has the ability to resist fracture propagation in gas pipelines. To establish these properties, a drop weight tear test (DWTT) is carried out. The average of two shear tests is specified to be $\geq$ 85%, at a given test temperature. The good shear value, in combination with acceptable CNV values, gives confidence in a material's ability to resist fracture.

More detail on the ductile fracture is discussed in Annex G of API 5L. Readers are advised to reference current versions of API 5L/ISO 3183 for more updated properties of line pipes.

API 5L has added specific normative annexes to address additional requirements for line pipes. They address the requirements for:

1. Jointers
2. Manufacturing procedure qualification for PSL 2 pipes
3. Treatment of surface imperfections
4. Repair welding procedure
5. Non-destructive inspection for other than sour service or offshore service
6. Requirements for couplings (PSL 1 pipes only)
7. PSL 2 pipe with resistance to ductile fracture propagation
8. PSL 2 pipe ordered for sour service
9. PSL 2 pipe ordered as "through the flowline" (TFL) pipe
10. PSL 2 pipe ordered for offshore service
11. Non-destructive inspection for pipe ordered for sour service and/or offshore service.

As the name suggests, these normative annexes are specifically added to address special requirements. There are four more annexes detailing information on:

1. Steel designations as used in Europe
2. Correspondence of terminology between ISO 3183 and its source documents

3. Identification and explanation of deviations
4. API monogram use and importance.

ORDERING A LINE PIPE

Section 7 of API 5L 44th edition has an extensive 55-point checklist of information that the purchaser should provide to the supplier. The purchaser can use this list to prepare the project specification and base the purchase order on the information.

In context with the above, it is known that most of the pipes are not bought from mills but from suppliers who have no part in manufacturing the pipe. In such situations, not much is in the control of the purchaser, except to keep looking for whatever is available on the market and try to best match their need with availability. However, that is not the best practice, though it is the most frequently followed.

PIPES FROM OTHER SPECIFICATIONS

Pipes from other specifications, especially ASTM or ASME, are commonly used for piping systems, especially small diameter and some larger diameter above ground piping in plants and facilities. These are often ASTM A-106, A-53, or A-333 pipes, pipes and equipment made from ASTM A 516 of various grades, ASTM A-537 Cl I, II, and III plates, etc. Where a substitution is made from API 5L pipes to ASTM pipes, the engineering team must evaluate substituted material based on the SMYS, heat treatment conditions, required impact test – including both temperature and energy absorption CVN values. It may be noted that some of the ASTM pipes and materials are not mandated to be impact tested, and when required, the impact energy CVN absorbed values for several ASTM/ASME materials are significantly lower than the minimum average 20 ft-lbs specified for API 5L pipes.

CHAPTER THREE

Fittings and Forgings

Contents

Pipeline Integrity Handbook
ISBN 978-0-12-387825-0

Fittings are a very essential part of any engineering construction; pipelines and piping are no exception. The term may cover elbows, tees, reducers, both eccentric and concentric, segmentable bends, induction bends, field bends, flanges, weldolets, anchor forgings, forged fittings, etc. These materials cover a large list of shapes and sizes and they also come in different material strength levels to match pipe material and service conditions. Various grades within the ASTM A-234 specification are specified for a number of wrought carbon and alloy steel fittings for piping. For reference only, the general properties of various grades are given in Table 3-3-1. It is strongly recommended that engineers and professionals procure the latest version of the referenced codes and specifications for more current and accurate information on materials.

A number of ASME and Manufacturers Standard Society (MSS) specifications that are commonly used in pipeline construction are discussed, with general information on materials that is useful in making correct decisions in selection of materials for specific project needs. MSS specifications MSS-SP 75 for High-Test, Wrought, Butt Welding Fittings and MSS-SP 44 Steel Pipeline Flanges are design and material specifications. They have their own standards for chemical and mechanical properties of material, but they also reference several ASTM specifications like ASTM A-105, A-106, A-53, A-234, A-420, and A-694 for material. Both include material with higher yield strength, as given in Table 3-3-2.

It may be noted that ASME A234, described above, is essentially a material specification, and specific fitting

Table 3-3-1 Chemical Compositions of Various Wrought Steel Fittings.

The carbon steel grades:

Grade	C	Mn	P	S	Si	Cr	Mo	Tensile × 1000 psi	Yield × 1000 psi
WPB	0.30 max	0.29–1.06	0.050 max	0.058 max	0.10 min			60–85	35
WPC	0.35 max	0.29–1.06	0.050 max	0.058 max	0.10 min			70–95	40

The following are the alloy steel grades:

Grade	C	Mn	P	S	Si	Cr	Mo	Ni	Cu	Tensile × 1000 psi	Yield × 1000 psi
WP1	0.28	0.30–0.90	0.045	0.045	0.10–0.50		0.44 -0.65			55–80	30
WP12 Cl 1 & Cl 2	0.05–0.20	0.30–0.80	0.045	0.045	0.60	0.80–1.25	0.44–065			70–95	40
WP11 Cl 1	0.05–0.15	0.30–0.60	0.030	0.030	0.50–1.00	1.00–1.50	0.44–0.65			60–85	30
WP11 Cl 2	0.05–0.20	0.30–0.80	0.04	0.04	0.50–1.00	1.00–1.50	0.44–0.65			70–95	40
WP11 Cl 3	0.05–0.20	0.30–0.80	0.04	0.04	0.50–1.00	1.00–1.50	0.44–0.65			75–100	45
WP22 Cl 1	0.05–0.15	0.30–0.60	0.04	0.04	0.50	1.90–2.60	0.87–1.13			60–85	30
WP22 Cl 3	0.05–0.15	0.30–0.60	0.04	0.04	0.50	1.90–2.60	0.87–1.13			75–100	45
WP 5	0.15	0.30–0.60	0.40	0.30	0.50	4.0–6.0	0.44–0.65			60–85	30
WP 9	0.15	0.30–0.60	0.03	0.03	0.25–1.00	8.0–10.00	0.90–1.10			60–85	30
WPR	0.20	0.40–1.06	0.045	0.050				1.60–2.24	0.75–1.25	63–88	46
WP 91	0.08–0.12	0.30–0.60	0.20	0.10	0.20–0.50	8.0–9.5	0.85–1.05	0.40	V, Co, N, Al	85–110	60

Single numbers are maximum values unless stated otherwise.

Table 3-3-2 Properties of Forging and Fitting Material Grades.

MSS-SP 75

Class symbol	Minimum yield strength (psi)	Minimum tensile strength (psi) for all thickness	Minimum elongation in 2 inch %	Maximum carbon equivalent $(CEq)_{IIW}$
WPHY-42	42,000	60,000	25	0.45
WPHY-46	46,000	63,000	25	0.45
WPHY-52	52,000	66,000	25	0.45
WPHY-56	56,000	71,000	20	0.45
WPHY-60	60,000	75,000	20	0.45
WPHY-65	65,000	77,000	20	0.45
WPHY-70	70,000	82,000	18	0.45

MSS-SP 44

Grade	Minimum yield strength (psi)	Minimum tensile strength (psi) for all thickness	Minimum elongation in 2 inch %	Maximum carbon equivalent $(CEq)_{IIW}$
F36	36,000	60,000	20	0.48
F42	42,000	60,000	20	0.48
F46	46,000	60,000	20	0.48
F48	48,000	62,000	20	0.48
F50	50,000	64,000	20	0.48
F52	52,000	66,000	20	0.48
F56	56,000	68,000	20	0.48
F60	60,000	75,000	20	0.48
F65	65,000	77,000	18	0.48
F70	70,000	80,000	18	0.48

dimensions and tolerances are governed by one of the ASME B16 specifications, discussed below.

Flanges and fittings of up to 24 inches are commonly covered under ASME B16.5 or 16.9. When the diameter exceeds 24 inches it may be necessary to reference MSS-SP 75 for fittings and MSS-SP 44 for flanges. MSS specifications also have alloy steel high yield strength material. Some of these materials are not ASME materials, however, and care must be taken to ensure that such materials meet the necessary calculations. The materials specified also differ in some grades of MSS-SP products. They also address the low temperature and impact energy absorption requirements. These fittings are often quenched and temper heat-treated.

The term "forgings," for this discussion, includes flanges, forged fittings, anchor forgings, weldolets, etc. These are heavy sections of material with different microstructures; they behave very differently in their cooling and heating cycles, which assumes a complex relationship when they are welded with other wrought material. Specifically in welding, the bulk of material works as a heat sink affecting the directional heat flow during welding. Due to the thickness of the material, the heat flow is often three-directional, raising the relative thickness values. This gives rise to the situation where simple and uniform cooling cannot be predicted, as the carbon equivalent (CEq) value of forgings is often higher than the CEq values specified for pipes to which they are welded. This demands that, when welding pipe to forging, a suitable welding procedure is developed to control cooling rates to avoid formation of harmful martensite and subsequent cracking. The higher alloyed forging specifications like ASTM A-694 and MSS-SP 44 specify the maximum CEq, and these are often higher than most of the wrought materials they are welded with. Engineers have the responsibility to assess and determine if they would specify some lower values as their acceptable maximum

CEq to suit what level is appropriate to meet the requirements of the work. This is especially important in pipeline construction where welding procedure specification (WPS) are qualified with limited pre-heating and without post-weld heat treatment. As a general guide, a CEq exceeding 0.39 should be avoided if proper pre-heat or post-weld heat treatment (PWHT), or both, cannot be included in the welding procedure. It may be added that CEq is a relative number, not a percentage, though many specifications and books erroneously report CEq as percentage (%).

In the following paragraphs, the related specifications are introduced and discussed briefly. For more detailed information, the relevant specification must be reviewed prior to making a decision.

ASME/ANSI B16.5 – PIPE FLANGES AND FLANGED FITTINGS

This standard for flanges and flanged fittings covers pressure-temperature ratings, materials, dimensions, tolerances, marking, testing, and methods of designating openings for pipe flanges and flanged fittings.

The standard includes flanges with rating class designations of 150, 300, 400, 600, 900, 1500, and 2500 in sizes NPS 1/2 through NPS 24. The requirements are given in both metric and US units. The standard is limited to flanges and flanged fittings made from cast or forged materials, as well as blind flanges and certain reducing flanges made from cast,

Table 3-3-3 Maximum Allowable Non-Shock Pressure *(psig).*

	Pressure class *(lb)*						
	150	**300**	**400**	**600**	**900**	**1500**	**2500**
	Hydrostatic test pressure (psig)						
Temperature (°F)	**450**	**1125**	**1500**	**2225**	**3350**	**5575**	**9275**
−20 to 100	285	740	990	1480	2220	3705	6170
200	260	675	900	1350	2025	3375	5625
300	230	655	875	1315	1970	3280	5470
400	200	635	845	1270	1900	3170	5280
500	170	600	800	1200	1795	2995	4990
600	140	550	730	1095	1640	2735	4560
650	125	535	715	1075	1610	2685	4475
700	110	535	710	1065	1600	2665	4440
750	95	505	670	1010	1510	2520	4200
800	80	410	550	825	1235	2060	3430
850	65	270	355	535	805	1340	2230
900	50	170	230	345	515	860	1430
950	35	105	140	205	310	515	860
1000	20	50	70	105	155	260	430

forged, or plate materials. Also included in this standard are requirements and recommendations regarding flange bolting, flange gaskets, and flange joints.

Flanges have temperature and pressure class ratings assigned to them. Table 3-3-3 gives, for reference only, the maximum non-shock pressure (psig) for pressure class ratings 150 to 2500. The dimensions of various classes ranging from Class 150 to Class 2500 ASME B16.5 are given, and readers are advised to reference these specifications for up-to-date correct information.

ASME/ANSI B16.9 – FACTORY-MADE WROUGHT STEEL BUTT WELDING FITTINGS

This standard covers overall dimensions, tolerances, ratings, testing, and markings for wrought factory-made butt welding fittings in sizes NPS 1/2 through 48 (DN 15 through 1200). Various shapes and sizes of fittings are manufactured to meet all possible engineering requirements, and they are available in different class ratings or by wall thickness classifications. The dimensions of various types of fittings listed below are given in ASME B16.9, and readers are encouraged to reference these tables for various critical dimensions for each of these fittings:

- Long radius elbows
- Long radius reducing elbows
- Long radius returns (U-bends)
- Short radius elbows
- Long radius returns
- 3D-90° and 45° elbows
- Tees and crosses
- Lap joint stub ends
- Caps
- Eccentric and concentric reducers.

The tolerance on the dimensions of the above listed fittings is detailed in a table in ASME B16.9.

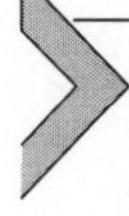

ASME/ANSI B16.11 – FORGED STEEL FITTINGS, SOCKET WELDING, AND THREADED CONNECTIONS

This standard covers ratings, dimensions, tolerances, marking and material requirements for forged fittings, both socket welding and threaded connections.

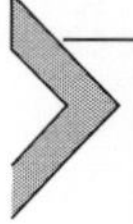

ASME/ANSI B16.14 – FERROUS PIPE PLUGS, BUSHINGS, AND LOCKNUTS WITH PIPE THREADS

This standard for ferrous pipe plugs, bushings, and locknuts with pipe threads addresses:

(a) Pressure-temperature ratings
(b) Size
(c) Marking
(d) Materials
(e) Dimensions and tolerances
(f) Threading
(g) Pattern taper.

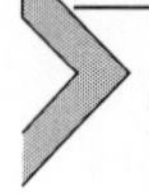

ASME/ANSI B16.20 – METALLIC GASKETS FOR PIPE FLANGES, RING-JOINT, SPIRAL-WOUND, AND JACKETED GASKETS

This standard covers materials, dimensions, tolerances, and markings for metal ring-joint gaskets, spiral-wound metal gaskets, and metal-jacketed gaskets and filler material. These gaskets are dimensionally suitable for use with flanges described in the reference flange standards ASME/ANSI B16.5, ASME B16.47, and API-6A. This standard covers spiral-wound metal gaskets and metal-jacketed gaskets for use with raised face and flat face flanges.

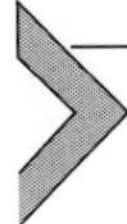

ASME/ANSI B16.21 – NONMETALLIC FLAT GASKETS FOR PIPE FLANGES

This standard is for nonmetallic flat gaskets for bolted flanged joints in piping and it includes the following:

(a) Types and sizes
(b) Materials
(c) Dimensions and allowable tolerances.

ASME/ANSI B16.25 – BUTT WELDING ENDS

This standard covers the preparation of butt welding ends of piping components to be joined into a piping system by welding. It includes requirements for welding bevels, for external and internal shaping of heavy-wall components, and for preparation of internal ends (including dimensions and tolerances). Coverage includes preparation for joints with the following:

(a) No backing rings
(b) Split or non-continuous backing rings
(c) Solid or continuous backing rings
(d) Consumable insert rings
(e) Gas tungsten arc welding (GTAW) of the root pass.

Details of preparation for any backing ring must be specified when ordering the component.

ASME/ANSI B16.28 – WROUGHT STEEL BUTT WELDING SHORT RADIUS ELBOWS AND RETURNS

This standard covers ratings, overall dimensions, testing, tolerances, and markings for wrought carbon and alloy steel butt welding short radius elbows and returns. The term wrought denotes fittings made of pipe, tubing, plate, or forgings.

ASME/ANSI B16.36 – ORIFICE FLANGES

This standard covers flanges (similar to those covered in ASME B16.5) that have orifice pressure differential connections. Coverage is limited to the following:

(a) Welding neck flanges Classes 300, 400, 600, 900, 1500, and 2500
(b) Slip-on and threaded Class 300
(c) Orifice, nozzle, and venturi flow rate meters.

ASME/ANSI B16.39 – MALLEABLE IRON THREADED PIPE UNIONS

This standard for threaded malleable iron unions, Classes 150, 250, and 300, provides requirements for the following:

(a) Design
(b) Pressure-temperature ratings
(c) Size

(d) Marking
(e) Materials
(f) Joints and seats
(g) Threads
(h) Hydrostatic strength
(i) Tensile strength
(j) Air pressure test
(k) Sampling
(l) Coatings
(m) Dimensions.

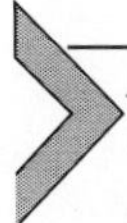

ASME/ANSI B16.42 – DUCTILE IRON PIPE FLANGES AND FLANGED FITTINGS, CLASSES 150 AND 300

This standard covers minimum requirements for Class 150 and 300 cast ductile iron pipe flanges and flanged fittings. The requirements covered are as follows:

(a) Pressure-temperature ratings
(b) Sizes and method of designating openings
(c) Marking
(d) Materials
(e) Dimensions and tolerances
(f) Blots, nuts, and gaskets
(g) Tests.

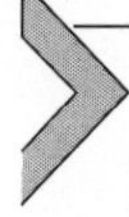

ASME/ANSI B16.47 – LARGE DIAMETER STEEL FLANGES: NPS 26 THROUGH NPS 60

This standard covers pressure-temperature ratings, materials, dimensions, tolerances, marking, and testing for pipe flanges in sizes NPS 26 through NPS 60 and in ratings Classes 75, 150, 300, 400, 600, and 900. Flanges may be cast, forged, or plate (for blind flanges only) materials. Requirements and recommendations regarding bolting and gaskets are also included.

ASME/ANSI B16.48 – STEEL LINE BLANKS

This standard covers pressure-temperature ratings, materials, dimensions, tolerances, marking, and testing for operating line blanks in sizes NPS 1/2 through NPS 24 for installation between ASME B16.5 flanges in the 150, 300, 600, 900, 1500, and 2500 pressure classes.

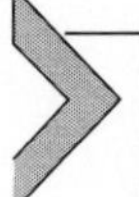

ASME/ANSI B16.49 – FACTORY-MADE WROUGHT STEEL BUTT WELDING INDUCTION BENDS FOR TRANSPORTATION AND DISTRIBUTION SYSTEMS

This standard covers design, material, manufacturing, testing, marking, and inspection requirements for factory-made pipeline bends of carbon steel materials having controlled chemistry and mechanical properties, produced by the induction bending process, with or without tangents. This standard covers induction bends for transportation and distribution piping applications (e.g., ASME B31.4, B31.8, and B31.11). Process and power piping have differing requirements and materials that may not be appropriate for the restrictions and examinations described herein, and therefore are not included in this standard.

Valves

Contents

Pipeline Integrity Handbook
ISBN 978-0-12-387825-0

Valves in oil and gas installations are governed by several standards and specifications that are issued by many organizations. They are dynamic documents that reflect sound engineering and change to the needs of the industry based on engineering practices, changes in market demands, and changes in technology and manufacturing procedures.

The valve standards focus on important aspects that include type of valves such as gate, globe, or check valves and materials.

All aspects of valve design, functionality, inspection, and testing are covered in dozens of ASME, API, and MSS documents, as well as various other international standards. The number of codes, standards, and specifications can make the procurement of valve products a job for experts only.

A good understanding of the primary standards affecting these products is necessary. It is recommended that current copies of all the valve standards that apply to the project are referenced.

A brief introduction to some of the specifications in common use is given below.

API 598 VALVE INSPECTION AND TESTING

This specification covers the testing and inspection requirements of gate, globe, ball, check, plug, and butterfly valves. The test pressure for each valve must be determined on the basis of the tables given in ASME/ANSI 16.34.

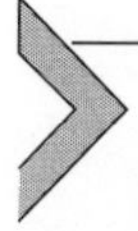

API 600 STEEL VALVES - FLANGED AND BUTT WELDED ENDS

API 600 and ISO 10434 are the primary steel gate valve specifications. They address design, construction, and testing criteria. An appendix in the API 600 specification addresses pressure seal valves. API 600 also lists important dimensions such as stem diameter minimums, wall thickness, and stuffing box size.

Another important gate valve specification is ASME/ANSI B16.34. This document gives extensive details on valves built to meet ASME boiler code pressure-temperature ratings. One important point of difference from API 600 to ASME/ANSI B16.34 is the wall thickness requirement. API 600 requires a heavier wall for a given pressure rating than does ASME/ANSI B16.34. This difference must be kept in mind when working with the two different codes.

API 602 COMPACT STEEL GATE VALVES - FLANGED, THREADED, WELDING AND EXTENDED-BODY ENDS

This specification addresses gate valve sizes 4 inch and smaller. Valves starting from Class 150 through Class 1500 are covered by API 602. This specification covers the same details for small forged gate valves that API 600 does for larger valves. API 602 further gives dimensions for extended-body valves, which are used extensively in industrial facilities.

Similar to API 600 above, API 602 also requires a heavier wall for Classes 150, 300 and 600 as compared with ASME/ANSI B16.34 requirements.

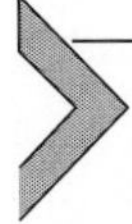

API 603 CLASS 150 CAST, CORROSION RESISTANT FLANGED-END GATE VALVES

This standard covers corrosion resistant bolted bonnet gate valves with flanged or butt welded ends in sizes NPS ½ inch to 24 inch, corresponding to nominal pipe sizes ASME B31.10M and Classes 150, 300, and 600 as specified in ASME B16.34.

API 608 METAL BALL VALVES - FLANGED AND BUTT WELDED ENDS

This specification covers Class 150 and 300 metal ball valves that have either butt welded or flanged ends and are for use in on–off service. It addresses the design and construction criteria. The important feature to note is that the working pressure for ball valves must be based on the seat material and not on the class of the valve.

These valves are commonly available in cast steel meeting ASTM A-351 grade CF8M and CF8; however, other corrosion resistant alloys meeting other specifications and grades are also used to manufacture these valves.

API 609 BUTTERFLY VALVES – DOUBLE FLANGED, LUG- AND WAFER-TYPE

These valves are designed to meet up to Class 600 rating and are intended to be fitted between flanges. The standard covers design, materials, face-to-face dimensions, pressure-temperature ratings, and examination, inspection, and test requirements for gray iron, ductile iron, bronze, steel, nickel-based alloy, or special alloy butterfly valves that provide tight shut-off in the closed position and are suitable for flow regulation.

TESTING OF VALVES

As described above, the test specification for most valves is API 598 "Valve Inspection & Test." Most metallic seated valves larger than ANSI size 2-inch have an allowable leakage rate that is listed in API 598. Some valve types such as bronze gate, globe, and check valves are usually not tested as per API 598. These are normally tested per MSS SP-61 "Pressure Testing of Steel Valves."

Pipeline valves are often specified to meet API 6D/ISO 14313 requirements. This document covers the design, materials, and dimensions of valves for pipeline service. The testing requirements of API 6D differ from those of API 598, the primary difference being API 6D's zero allowable leakage on closure tests. Since most of the valves built to API 6D are resilient seated, this is easily achieved. However, when the test standard is applied to metallic seated wedge gate, globe, or check valves, compliance can be difficult.

NACE TRIM AND NACE MATERIAL

NACE MR0175/ISO 15156 Standard material requirements for sulfide stress cracking resistant metallic materials for oilfield equipment

This NACE MR 0175/ISO 15156 standard is now an ANSI specification; it is no longer a recommended practice as it was in the recent past. MR0175 is in three parts and helps determine the severity level of sour environment to various materials. The specification does not list materials but guides on how to assess, and if required, to qualify materials for specific service conditions. Most of the materials are identified by their UNS numbers and fabrication techniques.

Designed to lessen the likelihood of H_2S induced cracking, in the valve industry often the term "NACE trim" is used; this simply means that the trim material will be compliant with NACE/ANSI MR0175/ISO 15156 and will meet the given service conditions as specified.

ASME CODES AND OTHER SPECIFICATIONS

Although ASME Section IX, *Welding and Brazing*, ASME Section VII, *Div 1 Pressure Vessels Design*, ASME

B31.8, *Gas Transmission and Distribution Piping Systems*, and ASME B31.4, *Pipeline Transportation System for Liquid Hydrocarbon and Other Liquids*, are not valve standards, they have a strong influence on valve design, manufacturing, testing, and application. In pipelines, these specifications are regularly specified for welding, design, manufacture, and application.

Several international organizations have also issued valve standards, including the British Standards (BS), International Standards Organization (ISO), and The Canadian Standards Association (CSA). Some of the relevant valve standards are BS 1873 and BS 5352 for globe valves, BS 6364 for cryogenic service valves, and BS 1868 and BS 5352 for steel check valves. These documents are excellent starting points for those needing guidance in these particular areas. Table 3-4-1 shows some of the common valve testing standards.

Table 3-4-1 Common Valve Testing Standards.

Specification	Valve types
API 598	All
API 6D	Pipeline
ANSI B16.34	All
MSS SP-61	All
MSS SP-67	Butterfly
MSS SP-68	Butterfly
MSS SP-70	Iron
MSS SP-80	Bronze
MSS SP-81	Stainless steel knife gate valves

VALVE MATERIALS

Trim material

The internal metal parts of the valve, such as the ball, stem, and metal seats or seat retainers, should be of the same nominal chemical composition as the shell and should have mechanical and corrosion resistance properties similar to those of the shell. The purchaser may also specify a higher quality trim material.

Tables 3-4-2, 3-4-3, 3-4-4, and 3-4-5 show some of the common pipeline materials for specific parts of valves.

Table 3-4-2 Ball Valve Material – Body and Tailpiece Material Code Material Service Insert Conditions.

Material specification	Material	Service conditions
ASTM A-105	Carbon steel	Good for normal working conditions for pipeline. However, though the pipelines will not see temperatures at about 427°C, it may be noted that upon prolonged exposure to temperatures above 427°C, the carbide phase of carbon steel may be converted to graphite.
ASTM A-352 LF2	Low temp.	For service from −46°C to 340°C. This material must be quenched and tempered to obtain tensile and impact properties needed at subzero temperatures.
ASTM A-182 F304	18Cr-8Ni	Good creep strength, corrosion and oxidation resistance when exposed to temperatures above 427°C.
ASTM A-182 F316	18Cr-10Ni-2Mo	Good creep strength, corrosion and oxidation resistance when exposed to temperatures above 427°C, and it is resistant to formulation of Σ phase.

Table 3-4-3 Fastener Materials.

	Normal service	NACE compliant	Low temperature	Corrosion resistant	Others
Stud	A-193 Gr. B7	A-193 Gr. B7M	A-320 Gr.L7	A-193 Gr.B8(M)	Monel
Nut	A-194 Gr .2H	A-194 Gr. 2HM	A-194 Gr.L4	A-194 Gr. 8(M)	Monel

Table 3-4-4 Stem Materials.

ASTM specification	Service conditions
A-276 420/410	Good for service up to 425°C where corrosion and oxidation are not factors.
A-105 ENP	Good for service up to 425°C where corrosion and oxidation are not factors.
A-747 17-4 PH	Very high tensile material. Often used when differential hardness is required due to its resistance to galling. Material has higher corrosion resistance as compared to straight chromium alloy steels.
A-182 F316	Good creep resistance, corrosion and oxidation resistance when exposed to temperatures above 427°C and it is resistant to formulation of Σ phase.

COMPONENT MATERIALS FOR NACE STANDARDS

The component materials for NACE standards include compliance body and tailpiece materials (Table 3-4-6), compliant ball materials (Table 3-4-7), and compliance stem materials (Table 3-4-8). There are also several other valve specifications that address specific needs.

Table 3-4-5 Ball Materials.

ASTM specification	Material	Service conditions
A-105(N) ENP	Carbon steel	For service up to 538°C where corrosion and oxidation are not a factor.
A-182 F304(L), A-351 CF8(3)	18Cr-8Ni	Good creep resistance properties, corrosion and oxidation resistance when exposed to temperatures above 427°C.
A-182 F316 A-351 CF8M	18Cr-10Ni-2Mo	Good creep resistance properties, corrosion and oxidation resistance when exposed to temperatures above 427°C and it is resistant to formation of Σ phase.
A-182 F316 A-351 CF3M	18Cr-10Ni-2Mo	Good creep resistance properties, corrosion and oxidation resistance when exposed to temperatures above 427°C and it is resistant to formation of Σ phase.

Table 3-4-6 NACE Compliance Body and Tailpiece Materials.

Material specification	Type of material	Application
ASTM A-105	Forged carbon steel with ENP coating	For valve sizes 2 inch and above for all NACE compliant valves. Thickness of electro less nickel plate (ENP) is often limited to 0.003 inch
A-182 F304/ F316	Stainless steel forging	For valve sizes 2 inch and above for all NACE compliant valves.

Table 3-4-7 NACE Compliance Ball Materials.

Material specification	Type of material	Application
ASTM A-105	Forged carbon steel with ENP coating	For valve sizes 2 inch and above for all NACE compliant valves. Thickness of ENP is often limited to 0.003 inch.
A-182 F304/ F316	Stainless steel forging	For valve sizes 2 inch and above for all NACE compliant valves.

Table 3-4-8 NACE Compliance Stem Materials.

Material type	Properties	Application
Carbon steel bar	Maximum hardness not to exceed 22 HRC	For all classes valve sizes 2 to 4 inch. Some larger sizes, above 6 inch in size.
Stainless steel bar	17-4 PH	For all corrosive service or low temperature service requirements.

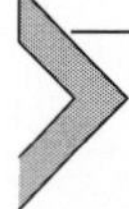

API 6D SPECIFICATION FOR PIPELINE VALVES

This API specification is possibly the most referenced valve specification in the pipeline industry. API Specification 6D is an adoption of ISO 14313: 1999, Petroleum and Natural Gas Industries – Pipeline Transportation Systems – Pipeline Valves. This international standard specifies requirements and gives recommendations for the design, manufacturing, testing, and documentation of ball, check, gate, and plug valves for application in pipeline systems.

Types of valves covered under this specification are gate valves, lubricated and non-lubricated plug valves, ball valves, and check valves with full or reduced opening configurations. The valves covered are from the range Class 150 (PN20) to Class 2500 (PN 420).

API 6FA

This specification is to establish the requirements for testing and evaluating the pressure containing performance of API 6D valves and also wellhead Christmas-tree equipment according to API 6A. The performance requirements are established regardless of the pressure rating or size.

The test covers the requirements of leakage through the valve and also the external leakage after exposure to fire with temperatures between 1400°F to 1800°F (761°C to 980°C) for a duration of 30 minutes.

API 526 FLANGED STEEL PRESSURE RELIEF VALVES

This standard addresses the basic requirements for direct spring-loaded pressure relief valves and pilot-operated pressure relief valves. This includes the orifice designation and area; valve size and pressure rating, inlet and outlet; materials; pressure-temperature limits; and center-to-face dimensions, inlet and outlet.

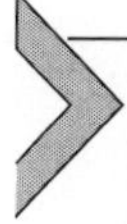

API 527 SEAT TIGHTNESS OF PRESSURE RELIEF VALVES (2002)

This API specification describes methods of determining the seat tightness of metal- and soft-seated pressure relief valves, including those of conventional, bellows, and pilot-operated designs.

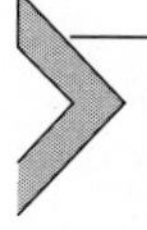

ANSI/API STD 594 CHECK VALVES – FLANGED, LUG, WAFER, AND BUTT WELDING

API Standard 594 covers design, material, face-to-face dimensions, pressure-temperature ratings, and examination, inspection, and test requirements for two types of check valves.

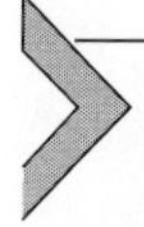

ANSI/API 599 METAL PLUG VALVES – FLANGED, THREADED, AND WELDING ENDS

This purchase specification covers requirements for metal plug valves with flanged or butt welding ends, as well as ductile iron plug valves with flanged ends, in sizes NPS 1 through NPS 24, which correspond to nominal pipe sizes in ASME B36.10 M. Valve bodies conforming to ASME B16.34 may have one flanged end and one butt welding end. It also covers both lubricated and non-lubricated valves that have two-way coaxial ports, and includes requirements for valves fitted with internal body, plug or port linings, or applied hard facings on the body, body ports, plug, or plug port.

ASME/ANSI B16.38 LARGE METALLIC VALVES FOR GAS DISTRIBUTION

This standard covers only manually operated metallic valves in nominal pipe sizes 2.5 inch to 12 inch that have the inlet and outlet on a common centerline. Theses valves are suitable for controlling the flow of gas from open to fully closed. These valves are often used in the distribution service lines, where the maximum guage pressure at which such distribution piping systems may be operated is directed by the regulatory bodies, such as the US Code of Federal Regulations (CFR), Title 49, Part 192, *Transportation of Natural and Other Gas by Pipeline*. The regulated maximum pressure in the USA is 125 psi (8.6 bar). Valve seats, seals, and stem packing may be nonmetallic.

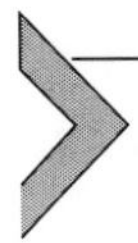

ASME/ANSI B16.33 MANUALLY OPERATED METALLIC GAS VALVES FOR USE IN GAS PIPING SYSTEMS UP TO 125 PSIG

This standard covers requirements for manually operated metallic valves, sizes NPS 1.2 through NPS 2, for outdoor installation as gas shut-off valves at the end of the gas service line and before the gas regulator and meter where the designated gauge pressure of the gas piping system does not exceed 125 psi (8.6 bar). The standard applies to valves operated in a temperature environment between −20°F and 150°F (−29°C and 66°C).

This standard sets forth the minimum capabilities, characteristics, and properties that a valve must possess at the time of manufacture, in order to be considered suitable for use in gas piping systems.

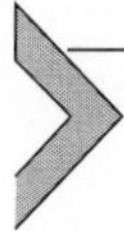

ASME/ANSI B16.40 MANUALLY OPERATED THERMOPLASTIC GAS VALVES

This standard covers manually operated thermoplastic valves in nominal sizes 1.2 through 6 These valves are suitable for use below ground in thermoplastic distribution mains and service lines. In the USA, the maximum pressure at which such distribution piping systems may be operated is in accordance with the CFR, Title 49, Part 192, *Transportation of Natural and Other Gas by Pipeline; Minimum Safety Standards*, for temperature ranges of −20°F to 100°F (−29°C to 38°C). This standard sets qualification requirements for each nominal valve size for each valve design as a necessary condition for demonstrating conformance to this standard. This standard sets requirements for newly manufactured valves for use in below ground piping systems for natural gas, including synthetic natural gas (SNG), and liquefied petroleum (LP) gases distributed as a vapor, with or without the admixture of air or as a mixture of gases.

ASME/ANSI B16.10 FACE-TO-FACE AND END-TO-END DIMENSIONS OF VALVES

This standard covers face-to-face and end-to-end dimensions of straightway valves, and center-to-face and center-to-end dimensions of angle valves. Its purpose is to assure installation interchangeability for valves of a given material, type size, rating class, and end connection.

ASME/ANSI B16.34 VALVES – FLANGED, THREADED, AND WELDING END

This standard applies to new valve construction and covers pressure-temperature ratings, dimensions, tolerances, materials, non-destructive examination requirements, testing, and marking. The specification covers cast, forged, and fabricated materials with flanged, threaded, and weld-ends, and wafer or flangeless valves of steel, nickel-based, and other alloys. Wafer or flangeless valves, bolted or through-bolt types, that are installed between flanges or against a flange are treated as flanged-end valves.

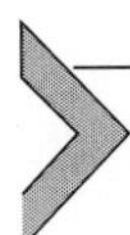

PREPARING A PIPING SPECIFICATION FOR A PROJECT

The engineering design will be meaningless if the concept of design is not effectively transferred to the field for correct execution. Not making effective communication of design choices of material will result in non-conformities, errors, and delay in execution. The effective way to control such errors and prevent escalation of project cost is to lay down a brief and easy to reference material class specification that would have extensive applicability in field operation. With such a specification in place, only very critical issues would be required to be referenced and decided at the project office. A typical piping specification prepared for Class 150 and Class 600 piping material for both above and below ground pipe would look like the example given below.

Definitions and abbreviations

a. BE = beveled ends
b. BW = butt weld
c. CS = carbon steel
d. CWP = cold working pressure
e. ERW = electric resistance welded
f. FBE = fusion bond epoxy
g. FO = full opening
h. GO = gear operated
i. HWO = hand wheel operated
j. NPS = nominal pipe size
k. PE = plain end
l. PN = pressure nominal
m. RFWN = raised face weld neck
n. SMLS = seamless
o. STD = standard
p. SW = socket weld
q. T&C = thread & coupled
r. THD = threaded
s. WE = weld end
t. XS = extra strong

PIPING CLASSIFICATIONS

Table 3-4-9 shows the four main piping classifications.

Table 3-4-9 Piping Classifications.

Class	Service	ANSI rating	Temperature
A1	Auxiliary and station piping (above ground)	150	−20° to 100°F
A2	Auxiliary and station piping (below ground)	150	−20° to 100°F
C1	Station piping (above ground)	600	−20° to 100°F
C2	Station piping (below ground)	600	−20° to 100°F

Class: A1 Above ground
Service: Auxiliary and station piping, for sweet crude oil
Pressure Rating: 285 psig
Temperature Rating: −20° to 100°F
Corrosion Allowance: None
Design Factor: 0.60
Design Code: ASME B31.4
Design Standard: N/A

Material specification

Table 3-4-10 shows material specification.

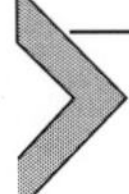

BRANCH CONNECTIONS

Refer to the branch connection chart.

Valves

Table 3-4-11 shows CS body, CS trim for gate, ball, globe, and check valves.

Table 3-4-12 shows the material specification.

Table 3-4-10 Material Specification.

Pipe	
NPS ½ to 1½	API 5L, Gr. B, SMLS, XS, T&C, Bare
NPS 2′	API 5L, Gr. B, SMLS, XS, BE, Bare
NPS 3 to 6	API 5L, Gr. B, SMLS STD, BE, Bare
NPS 8 to 24	API 5L, Gr. B, ERW, STD, BE, Bare
Flanges	
NPS ½ to 1½	NONE
NPS 2	CL. 150, RFWN, XS, ASTM A-105, ASME B16.5
NPS 3 to 24	CL. 150, RFWN, STD, ASTM A-105, ASME B16.5
Fittings	
NPS ½ to 1½	ASTM A-105, CL.2000, THD, ASME B16.11
NPS 2	ASTM A-234, Grade WPB, XS, BW, ASME B16.9
NPS 3 to 24	ASTM A-234, Grade WPB, STD, BW, ASME B16.9
Bolting	
ASTM A-193, Grade B7 c/w two (2) heavy hex nuts, ASTM A-194, Gr. 2H	
Gaskets	
ANSI Class 150 spiral wound gaskets per ASME B16.20, ⅛′ thick, non-asbestos filler, 304SS windings, carbon steel outer rings.	

BRANCH CONNECTIONS

Refer to the branch connection chart.

Table 3-4-13 shows CS Body, CS trim for gate, ball, globe, and check valves.

Table 3-4-14 shows a material specification.

Table 3-4-11 CS Body, CS Trim for Gate, Ball, Globe, and Check Valves.

Gate valves		
NPS 3/4 to 1½	Class 800 CS THD	API 602
NPS 2 to 6	Class 150 CS RF, Flex Wedge HWO	API 600
NPS 8 to 24	Class 150 F.O. CS RF, Slab GO	API 6D
Globe valves		
NPS 3/4 to 1½	Class 800 CS THD	ANSI B16.34
NPS 2 to 6	Class 150 CS RF, HWO	ANSI B16.34
NPS 8 to 12	Class 150 CS RF, GO	ANSI B16.34
Check valves		
NPS 3/4 to 1½	Class 800 CS Horizontal Piston THD	ANSI B16.34
NPS 2 to 24	Class 150 CS Dual Plate Wafer	ANSI B16.34
Ball valves		
NPS 3/4 to 1½	1500 PSI CWP (Min) CS THD	ANSI B16.34
NPS 2 to 6	Class 150 F.O. CS RF, Wrench	API 6D
NPS 8 to 24	Class 150 F.O. CS RF, GO	API 6D
Needle valves		
NPS 3/4 to 1	3000 PSI CWP (Min) CS THD	MSS SP-99
Gauge valves		
NPS 3/4 x ½	3000 PSI CWP (Min) CS,THD (M x F), MSS SP-99 w/bleeder	

Class: A2 Below ground
Service: Auxiliary and station piping, for sweet crude oil
Pressure Rating: 285 psig
Temperature Rating: −20° to 100°F
Corrosion Allowance: None
Design Factor: 0.60
Design Code: ASME B31.4
Design Standard: N/A

BRANCH CONNECTIONS

Refer to the branch connection chart.

Table 3-4-15 shows CS Body, CS trim for gate, ball, globe, and check valves.

Table 3-4-16 shows a material specification.

Table 3-4-12 Material Specification.

Material Specification	
Pipe	
NPS 2 to 6	API 5L, Gr. B, SMLS, XS, BE, c/w FBE
NPS 8 to 16	API 5L, Gr. B, ERW, STD, BE, c/w FBE
NPS 20 to 24	API 5L, Gr. B, ERW, STD, BE, c/w FBE
Flanges	Flanged connections in below ground piping is generally not specified, any use of underground flange connection must be reviewed and approved by engineering.
NPS 2 to 6	CL. 150, RFWN, XS, ASTM A-105, ASME B16.5
NPS 8 to 24	CL. 150, RFWN, STD, ASTM A-105, ASME B16.5
Fittings	
NPS 2 to 6	ASTM A-234, Grade WPB, XS, BW, ASME B16.9
NPS 8 to 24	ASTM A-234, Grade WPB, STD, BW, ASME B16.9
Bolting	
ASTM A-193, Grade B7 c/w two (2) heavy hex nuts, ASTM A-194, Gr. 2H	
Gaskets	
ANSI Class 150 spiral wound gaskets per ASME B16.20, ⅛′ thick, non-asbestos filler, 304SS windings, carbon steel outer rings.	

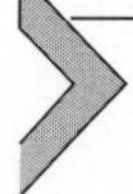

BRANCH CONNECTIONS

Table 3-4-17 shows …

BRANCH CONNECTION CHART

Table 3-4-18 shows an example of a branch connection chart.

Table 3-4-13 CS Body, CS Trim for Gate, Ball, Globe, and Check Valves.

Valves	
CS Body, CS trim for gate, ball, globe, and check valves.	
Gate valves	
NPS 2 to 6	Class 150 CS WE, Flex Wedge, HWO, API 600 w/Stem Ext
NPS 8 to 24	Class 150 CS WE, Slab, GO, w/Stem Ext; API 6D
Globe valves	
NPS 2 to 6	Class 150 CS WE, HWO, w/Stem Ext ANSI B16.34
NPS 8 to 12	Class 150 CS WE, GO, w/Stem Ext ANSI B16.34
Check valves	
NPS 2 to 24	Class 150 CS, WE, Swing Type API 6D
Ball valves	
NPS 2 to 6	Class 150 Stl WE, Wrench, w/Ext API 6D
NPS 8 to 24	Class 150 Stl WE, GO, w/Stem Ext API 6D
Needle valves	**None**
Gauge valves	**None**

Class: C1 Above ground
Service: Station piping, for sweet crude oil
Pressure Rating: 1480 psig
Temperature Rating: –20° to 100°F
Corrosion Allowance: None
Design Factor: 0.60
Design Code: ASME B31.4
Design Standard: N/A

Table 3-4-14 Material Specification.

	Material Specification
Pipe	
NPS ½ to 1½	API 5L, Gr. B, XS, SMLS, T&C, Bare
NPS 2	API 5L, Gr. B, XS, SMLS, BE, Bare
NPS 3 to 8	API 5L, Gr. B, STD, SMLS, BE, Bare
NPS 10 to 12	API 5L, Gr. B, XS, SMLS, BE, Bare
NPS 14 to 16	API 5L X-42, XS, ERW, BE, Bare
NPS 20	API 5L X-52, XS, DSAW, BE, Bare
NPS 24	API 5L X-60, XS, DSAW, BE, Bare
Flanges	
NPS 2	CL. 600 RFWN, XS, ASTM A-105, ASME B16.5
NPS 2 to 16	CL. 600, LAP JOINT, ASTM A-105, ASME B16.5 (Pump flanges only)
NPS 3 to 8	CL. 600 RFWN, STD, ASTM A-105, ASME B16.5
NPS 10 to 12	CL. 600 RFWN, XS, ASTM A-105, ASME B16.5
NPS 14 to 16	CL. 600 RFWN, XS, ASTM A-694 F42, MSS-SP44
NPS 20	CL. 600 RFWN, XS, ASTM A-694 F52, MSS-SP44
NPS 24	CL. 600 RFWN, XS, ASTM A-694 F60, MSS-SP44
Fittings	
NPS ¾ to 1½	ASTM A-105, CL 2000, THD., ASME B16.11
NPS 2	ASTM A-234 Gr. WPB, XS, BW, ASME B16.9
NPS 3 to 8	ASTM A-234 Gr. WPB, STD, BW, ASME B16.9
NPS 10 to 12	ASTM A-234 Gr. WPB, XS, BW, ASME B16.9
NPS 14 to 16	MSS SP-75, Gr WPHY42, XS, BW
NPS 20	MSS SP-75, Gr WPHY52, XS, BW
NPS 24	MSS SP-75, Gr WPHY60, XS, BW
Bolting	
ASTM A193 Grade B7 c/w two (2) heavy 2H hex nuts, ASTM A-194, Gr. 2H	
Gaskets	
ANSI Class 600 spiral wound gaskets per ASME B16.20, ⅛′ thick, non-asbestos filler, 304SS wound and carbon steel outer rings.	

Table 3-4-15 CS Body, CS Trim for Gate, Ball, Globe, and Check Valves.

Valves		
CS Body, CS trim for gate, ball, globe and check valves.		
Gate valves		
NPS 3/4 to 1½	Class 800 CS THD	API 602
NPS 2 to 6	Class 600 CS RF, Flex. Wedge, HWO	API 600
NPS 8 to 24	Class 600 CS RF, F.O., Slab, GO	API 6D
Globe valves		
NPS 3/4 to 1½	Class 800 CS THD	ANSI B16.34
NPS 2 to 4	Class 600 CS RF, HWO	ANSI B16.34
NPS 6 to 12	Class 600 CS RF, GO	ANSI B16.34
Check valves		
NPS 3/4 to 1½	Class 800 CS Horizontal Piston THD	ANSI B16.34
NPS 2 to 24	Class 600 F.O. CS Dual Plate Wafer	API 6D
Ball valves		
NPS 3/4 to 1½	1500 PSI CWP (Min) CS THD.	ANSI B16.34
NPS 2 to 6	Class 600 F.O. CS RF, Wrench	API 6D
NPS 8 to 24	Class 600 F.O. CS RF, GO	API 6D
Needle valves		
NPS 3/4 to 1	3000 PSI CWP (Min) CS THD	MSS SP-99
Gauge valves		
NPS 3/4 × ½	3000 PSI CWP (Min) CS, THD (M x F) w/bleeder	MSS SP-99

Class: C2 below ground
Service: Station piping, hydrocarbon liquid non-sour
Pressure Rating: 1480 psig
Temperature Rating: −20° to 100°F
Corrosion Allowance: None
Design Factor: 0.60
Design Code: ASME B31.4
Design Standard: N/A

Table 3-4-16 Material Specification.

	Material Specification
Pipe	
NPS 2 to 6	API 5L, Gr. B, XS, SMLS, BE, c/w FBE
NPS 8	API 5L, Gr. B, STD, SMLS, BE, c/w FBE
NPS 10 to 12	API 5L, Gr. B, XS, SMLS, BE, c/w FBE
NPS 14 to 16	API 5L X-42, XS, ERW, BE, c/w FBE
NPS 20	API 5L X-52, XS, ERW, BE, c/w FBE
NPS 24	API 5L X-60, XS, ERW, BE, c/w FBE
Flanges	Flanged connections in below ground piping is not specified, any use of underground flange connection must be reviewed and approved by engineering.
NPS 2 to 6	CL. 600 RFWN, XS, ASTM A-105, ASME B16.5
NPS 8	CL. 600 RFWN, STD, ASTM A-105, ASME B16.5
NPS 10 to 12	CL. 600 RFWN, XS, ASTM A-105, ASME B16.5
NPS 14 to 16	CL. 600 RFWN, XS, ASTM A-694 F42, MSS-SP44
NPS 20	CL. 600 RFWN, XS, ASTM A-694 F52, MSS-SP44
NPS 24	CL. 600 RFWN, XS, ASTM A-694 F60, MSS-SP44
Fittings	
NPS 2 to 6	ASTM A-234 Gr.WPB, XS, BW, ASME B16.9
NPS 8	ASTM A-234 Gr. WPB, STD, BW, ASME B16.9
NPS 10 to 12	ASTM A-234 Gr.WPB, XS, BW, ASME B16.9
NPS 14 to 16	MSS SP-75, Gr WPHY 42, XS, BW
NPS 20	MSS SP-75, Gr WPHY 52, XS, BW
NPS 24	MSS SP-75, Gr WPHY 60, XS, BW
Bolting	
ASTM A-193 Grade B7 c/w two (2) heavy 2H hex nuts, ASTM A-194, Gr. 2H	
Gaskets	
ANSI Class 600 spiral wound gaskets per ASME B16.20, ⅛′ thick, non-asbestos filler, 304SS windings, carbon steel outer rings.	

Table 3-4-17 CS Body, CS trim for gate, ball, globe, and check valves.

Valves		
Gate valves		
NPS 2 to 6	Class 600 CS WE, Flex. Wedge, HWO	API 600 w/Stem Ext.
NPS 8 to 24	Class 600 F.O. CS WE, Slab, GO	API 6D w/Stem Ext
Globe valves		
NPS 2 to 4	Class 600 CS WE, HWO w/Stem	ANSI B16.34 Ext.
NPS 6 to 12	Class 600 CS WE, GO w/ Stem Ext	ANSI B16.34
Check valves		
NPS 2 to 24	Class 600 F.O. WE Swing Type	API 6D
Ball valves		
NPS 2 to 6	Class 600 F.O. CS RF, Wrench w/Ext API 6D	
NPS 8 to 24	Class 600 F.O. CS WE, GO	API 6D w/Stem Ext.
Needle valves		
None		
Gauge valves None		

Table 3-4-18 Typical Branch Connection Chart.

Run																	
½"	3																
3/4"	3	3															
1"	3	3	3														
1½"	3	3	3	3													
2"	4	4	8	8	5												
3"	4	4	4	8	5	5											
4"	4	4	4	8	5	5	5										
6"	4	4	4	4	7	5	5	5									
8"	4	4	4	4	7	6	5	5	5								
10"	4	4	4	4	7	7	5	5	5	5							
12"	4	4	4	4	7	7	7	5	5	5	5						
14"	4	4	4	4	7	7	7	5	5	5	5	5					
16"	4	4	4	4	7	7	7	5	5	5	5	5	5				
20"	4	4	4	4	7	7	7	5	5	5	5	5	5	5			
24"	4	4	4	4	7	7	7	5	5	5	5	5	5	5	5		
30"	4	4	4	4	7	7	7	5	5	5	5	5	5	5	5	5	
36"	4	4	4	4	7	7	7	6	6	6	5	5	5	5	5	5	5
	½"	3/4"	1"	1½"	2"	3"	4"	6"	8"	10"	12"	14"	16"	20"	24"	30"	36"
	Branch																

1. (Not used)
2. (Not used)
3. Socket weld tee/reducing tee
4. Sockolet
5. Weld tee/reducing tee
6. Single outlet extruded header
7. Reinforcement contour saddle
8. Weld tee and concentric swage

Pressure Testing: Pneumatic and Hydro Test

Contents

Pipeline Integrity Handbook
ISBN 978-0-12-387825-0

PURPOSE

Pressure tests are carried out to induce a predetermined stress level, by pressurizing the equipment and observing the performance. Systems that are pressurized with air are often inspected with application of emulsified soap and inspected for rising bubbles from the leak locations. Some time pressurized components are immersed in water to detect leak locations.

Pressure testing, especially hydrostatic testing, has been used to determine and verify pipeline integrity. A truly vast amount of information can be obtained through this verification process; however, it is essential to identify the limits of the test process and obtainable results. There are also different types of pressure tests based on the objective of the test.

LEAK TESTING

Leak testing is carried out to determine the extent of a flaw's depth; that is, to ensure if the flaws in the material or weld extend to the surface of the material. The test can be either simple, pressurizing to a relatively low pressure to create a pressure differential and visually inspecting for leaks, or it can be more sophisticated, using electronic equipment such as acoustic emissions, to detect possible cracks and their growth over time.

Method

In its simplest form, the pressure test is carried out by increasing the internal pressure and creating a pressure differential at ambient pressure. This allows the liquid to flow out of possible openings, thereby identifying any leaks. These are visually inspected.

In cases where fluorescent dyes are used, the inspection is carried out in a dark place and black (ultraviolet) light is required. This kind of test is limited to smaller components and requires a specific setup; hence, it also has limitations within the premises of the manufacturing unit.

If the test media used is air or gas, then the leaks are located by looking for rising bubbles from the test surface, as an emulsion of mild soap and water is applied on the test surface.

Test media

The media used for leak testing can be any liquid that is not hazardous to personnel or the environment. Water is the most common medium for testing. Light oils are also used as test media. When water or any other liquid is used as a medium for testing, it is often called hydro testing. A typical pipeline hydro testing process is discussed at the end of this chapter.

Some tests require use of air or gas. Gas testing is very sensitive in detecting small leaks, but both air and gas as test media must be used with the utmost care, as they have the inherent disadvantage of explosive effects in the event of failures.

Sensitivity of the test

The test can be made more effective and sensitive by use of lighter chemicals, gases, or maybe by fluorescent dyes being added to the water.

The degree of sensitivity is adjusted with the required degree of flaw detection and relative degree of cost and risk involved in testing.

Proof testing

Proof testing is a relatively high-pressure test compared with the leak test. Proof testing is carried out to determine if the system can withstand applicable service loadings without failure or acquire permanent deformation in the part.

A proof test is generally designed to subject the material to stresses above those that the equipment is expected to carry in service. Such service stress is always below the yield strength of the material of construction.

The methods of these tests vary according to the specific design and requirements of the service. Mostly the proof test is applied in combination with visual inspection. These requirements are generally specified in the project specification or dictated by the code of construction.

The test medium is usually water (hydrostatic testing). If air is used as the medium (pneumatic testing), this is ordinarily restricted to relatively low-pressure testing because of the inherent safety issues associated with compressed air.

There are several types of flaws that can be detected by hydrostatic testing such as:

- Existing flaws in the material
- Stress corrosion cracking (SCC) and actual mechanical properties of the pipe
- Active corrosion cells
- Localized hard spots that may cause failure in the presence of hydrogen, etc.

There are some other flaws that cannot be detected by hydrostatic testing; for example, the sub-critical material flaws cannot be detected by hydro testing, but the test has a profound impact on the post-test behavior of these flaws.

The process of proof testing involves the following steps:

- Pressurizing the system with water to a stress level that is above the design pressure but below the yield strength of the material of construction
- Holding the pressure to the required time
- Monitoring for the drop in pressure
- Inspecting the object for any leak while it is under the test pressure
- Controlled depressurizing.

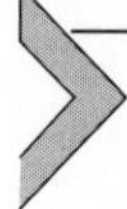

PRACTICAL APPLICATION OF HYDROSTATIC TESTING

Hydrostatic testing is used to determine and verify pipeline integrity. As stated earlier, there are several data that can be obtained from this verification process; however, it is essential to identify the limits of the test process and obtainable results. There are several types of flaws and information that can be detected by hydrostatic testing, like SCC, to determine the actual mechanical properties of the pipe.

Given that the test will play a significant role in the non-destructive evaluation of pipeline, it is important to utilize the test pressure judiciously.

The maximum test pressure should be so designed that it provides a sufficient gap between the test pressure and the maximum operating pressure (MAOP); in other words, the maximum test pressure should be greater than MAOP.

The determination of test pressure presupposes that after the test the surviving flaws in the pipeline shall not grow when the line is placed in service at the maintained operating pressure. For setting the maximum test pressure it is important to know two aspects of pressure on the existing defects in material:

- The immediate effect of pressure on the growth of defects leading to the failure of the test.
- The defects whose growth will be affected by pressure over time; these defects are often referred to as sub-critical defects because these will not fail during a one-time high-pressure test, but would fail at lower pressure if held for a longer time. The size of discontinuity that would be in the sub-critical group will be those that would fail independent of time at about 105% of the hold pressure. This implies that maximum test pressure would have to be

set to at least 5% to 10% above the MAOP in order to avoid growth of sub-critical discontinuities during the operational life of the pipeline.

Pressure reversal

The phenomenon of pressure reversal occurs when a defect survives a higher hydrostatic test pressure but it fails at lower pressure in a subsequent repressurization. One of several factors that work to bring about this phenomenon is the creep-like growth of sub-critical discontinuities over time at the lower pressure. The reduction in the wall thickness due to flaw growth in effect reduces the discontinuity depth. The material thickness to flaw depth (d/t) ratio reduces the ligament of the adjoining defects, which in effect reduces the required stress to propagate the discontinuity; this can be visualized as a chain reaction leading to failure. The other factor affecting the pressure reversal is the damage that occurs to the crack tip opening, as it is subject to some compressive force during initial pressure testing, leading the crack tip to force closed. The repressurization at a much lower pressure facilitates the growth of the crack. Hence, if such a pressure cycle is part of the design, then pressure reversal is a point to be considered in establishing the test pressure.

When a pipeline is designed to operate at a certain MAOP, it must be tested to ensure that it is structurally sound and can withstand safely the internal pressure before being put into service. Generally, pipelines are hydro tested by filling the test section of pipe with water and pumping the pressure up to a value that is higher than MAOP and holding it at this pressure for a period of 4 to 8 hours. The magnitude of the test pressure is specified by a code, and it is usually 125% of the operating pressure. Thus, a pipeline designed to operate continuously at 1000 psig will be hydrostatically tested to a minimum pressure of 1250 psig.

Let's consider a pipeline NPS 32, with 12.7 mm (0.500 inch) wall thickness, constructed of API 5L X70 pipe. Using a temperature de-rating factor of 1.00, we calculate the MAOP of this pipeline from the following:

$$P = \{2 \times t \times SMYS \times 1 \times factor(class1) \times 1\}/D \qquad [1]$$

Substituting the values:

$$2 \times 0.5 \times 70,000 \times 1 \times 0.72 \times 1/32 = 1575 \text{ psig}$$

For the same pipeline, if designed to a factor of 0.8, this pressure will be 1750 psig.

If the fittings are of ANSI Class 600, then the maximum test pressure will be $(1.25 \times 1440^{(\text{ASME B16.5})})$ 1800 psig.

If, however, ANSI 900 fittings were chosen, the pressure would be determined as below:

$$\left(1.25 \times 2220^{(\text{ASME B16.5})}\right) = 2775 \text{ psig.}$$

If the selected design factor is for Class 1 location, then the factor would be 0.72, as calculated above; in this case, the test would result in the hoop stress reaching 72% of the SMYS of the material. Testing at 125% of MAOP will result in the hoop stress in the pipe at a value of $1.25 \times 0.72 = 0.90$, or

90% of SMYS. Thus, hydro testing the pipe at 1.25 times the operating pressure, we are stressing the pipe material to 90% of its yield strength; that is 50,400 psi (factor 0.72 × 70 000).

If, however, we use a design factor of 0.8, as is now often used and is allowed by several industry codes, testing at 125% of MAOP will result in the hoop stress in the pipe of 1.25 × 0.8 = 1. In this case, the hoop stress would reach 100% of SMYS. So, at the test pressure of 1800 psig, the S_h (hoop stress) will be 56,000 psi (0.8 × 70,000). This will be acceptable in the case of Class 600, as fittings are the limiting factor.

But if the Class 900 fittings were the limiting factor of the system, then the maximum test pressure would be (1.25 × 2220) = 2775 psig, the resulting stress would be 88,800 psi, which will be far above the maximum yield stress of API 5L X70 PSL-2 material.

Critical flaw size

Using a test pressure at 100% of the material yield strength or above has some important pre-conditions attached to it. Such test pressure would require that the acceptable defect size in material and weld is reassessed. All other design conditions being equal, a higher design factor, resulting in thinner wall (d/t), will lead to reduction in the critical dimensions of both surface and through-wall defects. Critical surface flaw sizes, at design factors of 0.80 and 0.72, are stress dependent. Additionally, they will also be dependent on the acceptable Charpy energy for the material and weld.

As stated above, this increase in the d/t ratio in effect reduces the ligament of the adjoining defects, which progressively reduces the required stress to further propagate the discontinuity.

Note that flaws deeper than about 70% of wall thickness will fail as stable leaks in both cases. This statement implies that mere radiography of the pipe welds (both field and mill welds) will not suffice; automatic ultrasonic test (AUT) of the welds will be better suited to properly determine the size of the planer defects in the welds. Similarly, the use of AUT for assessing the flaws in the pipe body will have to be more stringent than usual.

Thinning due to corrosion

The wall thickness reduction due to corrosion is a complex phenomenon. The challenge it presents is unique and hence requires some additional steps to evaluate and take mitigative action. Thinning occurring due to general corrosion may be easier to assess and remedy than thinning associated with localized corrosion, also called "pitting." Often ultrasonic test (UT) or the magnetic flux leakage method is used combined with "smart pigging" to assess the wall thickness of the material.

Also of note is the fact that two or more areas of corrosion and metal loss separated by an area of sound wall may interact in such a manner that the effective reduction in wall may be more serious that one local pit (Figure 3-5-1). ASME B31.4 graphically describes two such scenarios:

1. In the first case, if the circumferential separation distance C is equal to, or greater than, 6 times the design wall

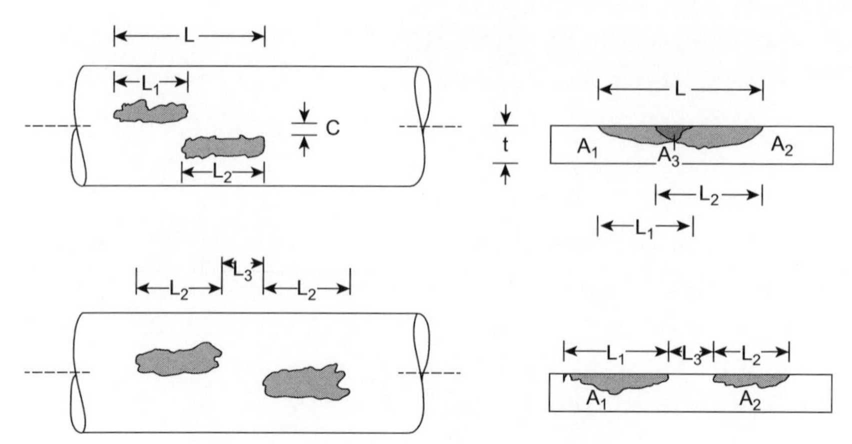

Figure 3-5-1 Defect sizing and interaction.

thickness, then both areas must be evaluated individually. If, on the other hand, the circumferential distance is less than 6 times the design wall thickness, then a composite area and length should be used for evaluation.

2. In the second case, the axial separation L is taken as the determining factor. If the distance L is greater than, or equal to, 1 inch (25.4 mm), then the two areas must be evaluated as separate anomalies. If axial separation is less than 1 inch (25.4 mm) then the two areas must be considered as one anomaly and, for that, the areas must be combined and the length L must also be added to make one anomaly.

Corrosion and Corrosion Protection

Contents

Pipeline Integrity Handbook
ISBN 978-0-12-387825-0

CORROSION

Corrosion is the deterioration of a material that results from a reaction with its environment. For metal in contact with an aqueous solution, the reaction is an electrochemical one involving the transfer of electrical charge (electrons) across the metal–solution interface.

The energy that exists in metals and causes them to corrode spontaneously results from the process of converting ore to metal. A measure of energy available in a metal, called Gibbs free energy, is required to power the corrosion reaction. This energy induced in the metal during the refining process is available as the potential energy (Δ-G^o) to power the reaction when metal is placed in an aqueous environment.

For a metal atom to leave the crystal structure, it must overcome the bonding energy with adjacent atoms in the crystal matrix. Metal atoms vibrate at their positions, and this vibration energy is dependent on the temperature. At ambient temperature, surface atoms have a better chance of leaving the crystal structure. This is possible because the atoms on the surface have fewer interatomic bonds than the internal atoms. In such situations, the vibrations associated with temperature may be sufficient for some atoms to escape the lattice structure, leaving behind some of its bonding electrons (ne^-) in an oxidation reaction:

$$M^o \rightarrow M^{n+} + ne^-$$

The oxidation process makes atoms more electropositive. Thus, it is now a metal ion, with a net positive charge, taking with it most of the atomic mass residing primarily in the nucleus. It is possible for the metal ion to return to the atom crystal structure; this means that the reaction is reversible. When a metal is placed in an aqueous solution, other possibilities for metal ions arise because of the presence of polar water molecules and other cations. Since water has polar molecules, it is attracted to the metal interface. A small number of water molecules will ionize to produce hydrogen ions (H^+) and hydroxyl ions (OH^-). The metal can now react anodically in two ways other than the one reaction discussed above. They can either produce metal hydroxide or aqueous ion, as the following reactions establish:

$$M^o + nH_2O \rightarrow M(OH)_n + nH^+ + ne^-$$

or

$$M^o + nH_2O \rightarrow {MO_m}^{n-2m} + 2mH^+ + ne^-$$

In each of the reactions, positive charges on the solution side of the interface are produced, and left behind are the negative charges in the metal. In the aqueous solution this will create a potential difference (E) across the interface. This potential is a function of the metal involved and the pH of the electrolyte. This tendency of metal to corrode by one of the three reactions stated above can be calculated using energy relationships. For iron a potential-pH diagram is plotted. The three diagrams below depict these conditions, including the Evans Corrosion Cell diagram (Figure 3-6-1), which shows the gap between cathode polarization and anode polarization. It is this gap that needs to be bridged by a cathodic protection (CP) system to stop corrosion. In the

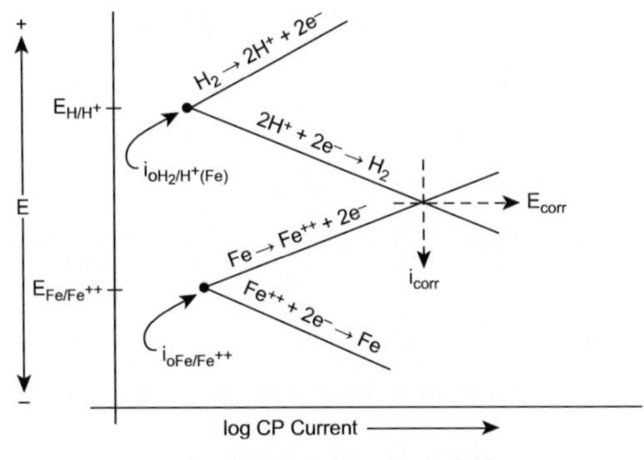

a - Polarization Curve – Iron In Acid

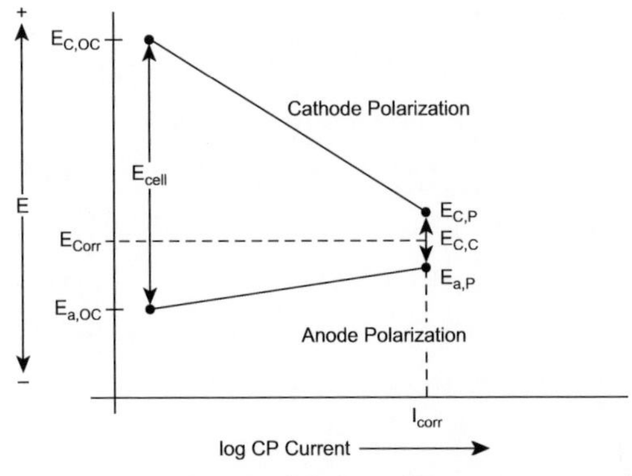

b - Corrosion Cell – Evans Diagram

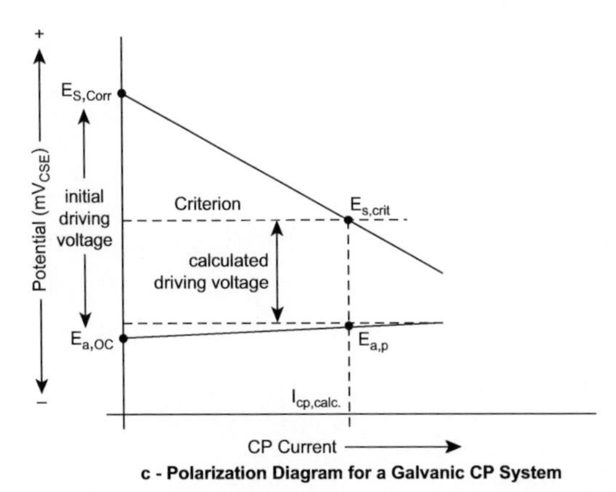

c - Polarization Diagram for a Galvanic CP System

Figure 3-6-1 Corrosion potentials and polarization diagram.

first diagram (a) the corrosion polarization of iron in an acidic environment is shown. Note the evolution of hydrogen and that the polarization takes place at the i_{corr} and E_{corr} interaction point. The diagram (c) is a practical application of the Evans diagram depicting the galvanic CP system.

Protection of assets from corrosion and resulting damages and losses, in terms of environment, material, and sometimes lives, has assumed center stage in the industry. The entire concept of integrity management revolves around this basic concept. However, this basic concept is not so basic after all. It involves knowledge of various modes, methods, and conditions of corrosion and designing the system to eliminate, reduce, or control corrosion to practical and manageable levels. The corrosion protection measures are discussed in subsequent paragraphs; however, it is important to understand some of the steps that must be taken at the design stage to avoid the majority of corrosion situations.

Control of corrosion involves several stages that begin with a potential corrosion study and then address the protection methods. All of the following must be considered and addressed:

- **Selection of material** – Selection of the correct material for the system is the primary step in control of corrosion. As we will see, all the other steps come back to this point.
- **Process parameters** – This will involve temperature, velocity of flow, type of flow, pressure, and chemistry of the environment.

- **Operating lifetime** – The design life of the system must be determined and the protection and design must be provided for the design life.
- **Construction parameters** – This will include stresses caused during various construction activities, like welding design and process quality control, lowering in trenches, soil type, elevations, etc.
- **Corrosion allowance** – Where it is necessary, adequate corrosion allowance to the material must be provided. Often this allowance varies from 0.0625 inch (1.6 mm) to 0.125 inch (3 mm); however, more than a 0.125-inch (3 mm) allowance is also provided in exceptionally corrosive systems. When the system is too corroded, alternative corrosion protection measures are required, that may include special coatings, lining, or even use of corrosion resistant alloys (CRAs). These are often clad on the carbon manganese substrate of the exposed surface, often on the inside of the pipeline.
- **Drainage geometry –** Accumulation of fluid and corrosive medium contributes to corrosion damages. Design should avoid dead ends and provide for proper flow of and treatment of drain system.
- **Dissimilar metals –** Use of dissimilar metal causes galvanic corrosion. Knowledge of galvanic potential of materials and the Nernst equation to understand corrosion potential can be applied; this is helpful in this stage of design:

$$E = E^\circ - (RT/nF) \times \text{In}\ [\text{reaction products/reactants}]$$

Where:
E = Actual reaction potential
E° = Potential under standard conditions

R = Gas constant
T = Temperature in Kelvin
n = Number of electrons transferred in reaction
F = Faraday's constant (96,500 coul/mol)
In = Natural logarithm.

This can also contribute to corrosion in the presence of electrolyte. Avoiding such contact must be the first priority of engineering design. Using isolation and insulation between dissimilar metal is the solution. Isolation kits of various specifications are available on the market and suppliers' expertise must be obtained in selection of isolation and insulation materials suitable for specific environments.

- **Crevices** – Crevice corrosion is a form of localized corrosion attack in which the site of attack is a crevice, where free access to the surrounding environment is restricted. This could be oxygen concentration cell corrosion, in which the differential potential is caused by the different oxygen concentration in areas surrounding the crevice and in the crevice. In metal ion concentration cell corrosion the difference in potential between the inside and outside of the crevice is caused by a difference in metal ion concentration. The Nernst equation can be used to better understand concentration cell corrosion. The comparison of potentials at high and low concentration with standard potentials can be calculated. Crevice corrosion can be between metal-to-metal or metal-to-nonmetals, under deposits of debris. Design should take account of such situations and avoid creation of crevices.
- **Maintenance and inspection requirements** – Providing the access for maintenance and inspection is an essential step in corrosion control. Establishing a well planned

inspection and maintenance schedule is an essential part of a good integrity management system.

- **Corrosion protection** – The protection of external corrosion is provided primarily by a combination of coating and CP. They often complement each other.

PROTECTION FROM INTERNAL CORROSION

Internal corrosion is protected by control of the internal environment, which includes modifying the internal environment by use of various types of inhibitors, providing corrosion allowances, and establishing various monitoring devices. Internal coating, lining, and cladding are also used effectively in various environments; their use is, however, not general and is limited to specific environments. The other limiting factors are the cost-specific environment and practical application methods.

MONITORING OF INTERNAL CORROSION

The use of corrosion control methods described above has limited effect. It is always a good practice to establish a second type of system to know and understand what is going on inside the stem before it fails one day. The monitoring involves techniques ranging from simple visual inspection to very high tech electronic devices to monitor online and from remote locations. They include the following:

- Visual inspection
- NDE techniques – radiography, ultrasonic inspections, electromagnetic (Eddy current, magnetic flux leakage (MFL) inspections, liquid penetrant, magnetic particle testing)

- Specimen exposure (coupons and measurement of weight loss methods)
- Electrochemical (liner polarization resistance, electrochemical impedance spectroscopy, electrochemical noise)
- Water chemical analysis (determination of pH, chlorides, oxygen content, carbonates and bicarbonates, metal ions, etc.)
- Deposit analysis (suspended solids, scales, microbiological fouling, etc.)
- Monitoring the CP system.

APPENDIX

Internal Corrosion Monitoring Techniques for Pipeline Systems

Gerald K. Brown
Brown Corrosion Services, Inc., Houston, TX, U.S.A.
The Second Annual Internal Corrosion and Pipe Protection Conference
Sponsored by: Systems Integrity and Pipeline and Gas Journal, September 18–20, 1995
The Marriott Westside, Houston, TX, U.S.A.

ABSTRACT

The measurement of internal corrosion, and the methods used to obtain and gather these measurements, have often left the operators of internal corrosion monitoring systems frustrated because of the complicated and often difficult procedures involved. The current intrusive monitoring methods and the first generation of patch or external type devices have increased the cost of internal corrosion monitoring programs. In addition to the increasing cost, the true measurement of the actual internal corrosion has continued to evade those who need this data.

The latest developments of equipment, instrumentation, and software for the measurement of internal corrosion and internal corrosion control, now offer the operators an accurate and user-friendly alternative to what had been previously used. The instrumentation and software, by bringing in other parameters that cause the onset or increase in corrosion, now allow the operators to relate corrosion to the events that cause corrosion.

In this paper, I will discuss the cost of corrosion and corrosion control, the causes of failure, the causes of internal corrosion, and the different types of corrosion monitoring technology as well as recent examples of field data.

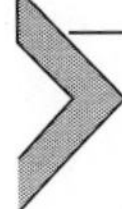

INTERNAL CORROSION OF A PIPELINE – MAIN CAUSES OF PIPELINE FAILURES

The Energy Resources Conservation Board (ERCB) in Alberta, Canada, (the petroleum governing body), published a breakdown of the main causes of pipeline failures in 1991 that covered a ten-year period. After taking into account the different uses of a pipeline (water, sour gas, natural gas, crude oil, and multiphase), corrosion accounted for the largest percentage of pipeline failures. The data shows that water lines represent 42% of the pipelines in Alberta and 72% of the failures associated with these water lines were caused by corrosion.

The ERCB also compiled and published a very interesting chart showing the causes of pipeline failures. This chart shows that the failures associated with external corrosion, welds, and equipment failures have remained fairly constant over this ten-year period. This finding tells us that the CP and coating industries have done a very good job of protecting the outside of the pipes and vessels from external corrosion related failures. However, by doing a good job externally, the pipelines have therefore lasted longer and the failures have shifted to those related to the internal area, and, as the chart shows, the failure rate because of internal corrosion is going off the scale. It is estimated that in Alberta alone, 80% of all pipeline corrosion failures are related to internal corrosion. This 80% of all failures rate is in spite of the internal corrosion monitoring and internal corrosion corrective measures that have been ongoing.

While the incidents of internal corrosion related failures in pipelines are high, the percentage in other sectors, such as the process industry, are just as high. While some license must be taken in averaging different industries, it is generally felt that a very high percentage of all corrosion failures across all industries is internal corrosion related. We feel that this figure represents all corrosion across the petroleum industry.

All of the failures contribute greatly to the failing infrastructure of the petroleum industry. The petroleum industry can classify all of these failures into two areas, and these two areas of failure are:

1. Mechanical failures.
2. Electrochemical (corrosion) failures.

Mechanical failures can be caused by many different phenomena including erosion, operational mistakes, operating at unsafe pressures, land subsidence, improper sizing of equipment, operational pressure and velocity changes, earthquakes, impacts, or wear. Electrochemical failures are caused by corrosion of metals and seals.

It is estimated that, of the approximate 20 billion dollars that are spent annually on corrosion related problems, approximately 65% of this cost could have been prevented. As an industry we have the process knowledge, the chemical inhibitor technology expertise, and other corrosion control methodologies to prevent approximately 12 billion dollars worth of damage, and to date we have not done enough to stop the destruction of our infrastructure in the petroleum industry.

The main reason that this damage occurs is because the ongoing corrosion process was not detected until the damage had already occurred. As an example, we have installed millions of dollars worth of intrusive internal corrosion monitoring equipment in pipelines where we have convenient access to the pipeline. Since corrosion often does not occur where we have the convenience of access, corrosion continues in other spots within the system where we are not monitoring. One of the main reasons we place monitors in convenient locations is the fact that the sensors have to be changed periodically. The same can be said of oil and gas production, the process industry, and most other phases of the petroleum sector.

In spite of the millions we spend on internal corrosion monitoring, the sensors are often not located where corrosion is occurring and it is not possible to locate them where corrosion does occur. In addition the actual corrosion mechanism and the corrosion rates causing many of the failures have not been apparent from the sensors that have been used. The sensors are often manufactured from different materials than that of the structure to be monitored and have been located in areas not similar to the corroding surfaces of the pipe and/or vessel.

CAUSES AND RATES OF INTERNAL CORROSION

The causes of internal corrosion are many and can be generally divided into the following areas:

1. The chemical composition of the stream.
2. The physical factors of the stream.
3. The physical factors of the structure.

1. Chemical composition of the stream – factors that affect the corrosion are:
   - H_2O content
   - H_2S content
   - CO_2 content
   - Dissolved solids
   - Organic and inorganic acids
   - Elemental sulfur and sulfur compounds
   - Bacteria and its by-products
   - Hydrocarbons
   - pH
   - Interactions of all of the above, other trace elements, and chemistry variables.
2. Physical factors of the stream – factors that affect the corrosion are:
   - Temperature
   - Pressure
   - Velocity
   - Vibration
   - Entrained solids and liquids
   - Deposits
   - Flow characteristics and patterns (slugs)
   - Interaction of all of the above and other physical factors of the stream.
3. Physical factors of the structure – factors that affect the corrosion are:
   - Materials of construction
   - Stresses: residual / operating
   - Design factors

- Crevices
- Depositions
- Surfaces
- Interactions of all of the above and other physical facets of the structure.

Most importantly, however, the interactions of all of these above factors can cause corrosion. As you can see with the interaction of all of these factors, plus others, the ability to predict the corrosion rate or the hydrogen permeation rate becomes very difficult at best.

Dynamic systems, by definition, are always changing and, therefore, corrosion types, rates, and locations change. The question is often asked as to why do the systems change, and why do we not stabilize the system variables and, thus, be able to control the corrosion and therefore, also control the internally related corrosion failures. The simple answers go back to the inlet products that are used, and the fact that different feed stocks must be used, which all have different characteristics. In addition, the flow from the original oil and gas wells change over time as the water cut, chemistry, pressure, and temperature changes occur during the depletion of the reservoirs that hold these fluids and gases. Of equal importance is the fact that to operate cost effectively today, different products must be placed in the pipelines to economically justify the use of existing facilities and the building of new facilities.

In pipeline operations the product also changes over time. In addition, the economics of running pipelines now call for transporting several different products through the pipelines.

By mixing products new situations arise where corrosion rates can be greatly increased. We are also faced with the situation where pipelines may be in the ground so long that slow corrosion rates can cause failures that we have not previously seen.

For all practical purposes, there will be no corrosion without water. Water, if only a thin film, is found in most petroleum-based systems throughout the world. In addition the sporadic wetting and drying that takes place during slug flow often exacerbates the corrosion mechanism. Water collection at certain parts of piping systems also leads to isolated corrosion cells.

METHODS OF CORROSION CONTROL

All methods of corrosion control fall into the following categories:

1. Coatings
2. CP (anodic protection)
3. Filming inhibitors
4. Alteration of the environment
5. Material selection
6. Alteration of the structure
7. Repair or replace

However, whatever method, or combination of methods, of corrosion controls one uses, corrosion monitoring must be in place. As an example, if you are using coatings for corrosion protection, when do you know if the coatings have failed, or how do you determine if the coatings are still protecting the surface? If the environment is altered by raising the temperature, how does one know if he has approached the onset or threshold of corrosion? If a pipeline dead leg is present in the system, how does one know if the corrosion in the dead leg is non-existent or accelerated? Internal monitoring can give you all of the answers.

Monitoring versus inspection

Inspection is a very important and necessary tool. However, any inspection program or inspection project, regardless of the method used, will only give you data on the metal deterioration between two or more specific points in time when the inspection was performed. Also, of most importance is the fact that all the data that is gathered is from the past. This means that if corrosion control measures are implemented, for example, increased doses of inhibitor or removal of water, the basis of choosing these actions is based upon the parameters as they were in the past, not as they are when the corrosion control measures are implemented. Also, the results of these corrosion control measures will not be observable until the next time of inspection.

By basing your corrosion control measures upon inspection rather than monitoring, you may not be able to actually stop or lower the corrosion rates which are the most common reason to monitor in the first place.

Monitoring, on the other hand, on a near real-time basis, will give you data on what is happening now.

By getting near real-time data you are able to relate corrosion data to events that caused the corrosion and therefore be able to lower corrosion rates by adjusting or eliminating the events that caused the corrosion to occur. Examples of obvious events causing corrosion could be the shutdown of a chemical injection pump, wash water escaping into the system, or a leaking pump sucking air (oxygen) into a closed liquid system. Once the upset condition is correlated with the increase in corrosion rate, these events can be stopped or modified so that they do not cause corrosion.

Inspection, however, should not be overlooked, but used in conjunction with monitoring. One of the most advantageous factors when using inspection techniques such as intelligent pigging is that this type of inspection can cover virtually whole piping systems. Monitoring does not cover the whole system and even the best engineering can miss the actual spots where corrosion may be occurring.

When the inspectors complete a project and leave the site the question remains; who is monitoring the system? A well planned monitoring program should be your sentinel until the next inspection.

Corrosion monitoring techniques

Several basic corrosion monitoring techniques have been used for many years. All of these methods have their place and work well in the proper locations and applications. Most of these traditional methods require intrusion into the stream to be monitored. This intrusion is necessitated because the sensor must be exposed to the environment to be studied. By making this intrusion, four general considerations should be of concern:

1. A hole of some sort must be made in the pipe and/or vessel.
2. The sensor is not of the same material as the structure to be monitored, nor is it in the exact same location where corrosion is occurring.
3. A probe changes flow patterns that, in turn, can determine different corrosion rates.
4. The surface areas of the sensors are restricted to very small areas.

The traditional monitoring methods have undergone drastic changes over the last several years due to the ability the computer has given us to digest large amounts of data and display this data in such a way that it can be correlated to the actual events that cause the corrosion. The uses of more sophisticated instrumentation and data loggers have allowed the users to 'look inside of their pipes.'

With the newer instrumentation and software, a whole range of innovative non-intrusive internal corrosion monitoring devices are now in the marketplace.

Major corrosion monitoring techniques include:

1. Coupons (C)
2. Electrical resistance (ER)
3. Linear polarization resistance (LPR)
4. Galvanic (ZRA)
5. Hydrogen probes and patches (H_2)

Several newer corrosion monitoring techniques are being tested in the laboratory, and in the field, and may prove of interest in the future. They are:

1. Hydrogen patches (H_2)
2. Impedance (EIS)
3. Electrochemical noise (ECN)
4. Surface resistance readings
5. Others (i.e., radiography, acoustic emission, etc.)

Weight loss corrosion coupons

Weight loss corrosion coupons are probably the oldest and still the most widely used method of corrosion monitoring. Coupons are simply a specimen of metal that is firstly weighed, then exposed into a specific environment, removed, and then cleaned and reweighed to determine the amount of weight loss corrosion that has occurred over a specific period of time. NACE has a standard method of determining the weight loss and the result of this determination gives one the amount of weight loss in mils per year (MPY) which is the most widely used reference. This reference is also used on most other corrosion monitoring methods as well. There are several different measurements in use, but MPY is by far the most widely used.

Weight loss corrosion coupons are available in many different shapes and materials. The size is not that important once one realizes that the more surface area exposed to the environment the more accurate will be the readings and the more quickly a reading can be obtained. The most common shape is called a strip coupon and is most often available in the following sizes; ½′ × 1⁄16′ × 3′ or ⅞′ × ⅛′ × 3′. Other configurations are also used such as rod or pencil coupons, pre-stressed coupons of several shapes, and 'lifesaver' shaped coupons that are often fitted with electronic probes.

Weight loss corrosion coupons are available in any material desired. The material is most often, however, either that of the material containing the environment (i.e., the pipe, vessel or tank) or very low carbon steel that is very susceptible to corrosion. If you are trying to duplicate the corrosion that is occurring on the pipe, you should use coupons of the same material. If you are using the coupons to determine if the inhibitor is filming and remaining in place, a low carbon steel coupon can be used. However, in using low carbon coupons, it must be remembered that if you can stop corrosion on these coupons, the corrosion rates of the pipe itself will not necessarily be affected.

A third use of coupons is for material selection. If a pipe, vessel or tank is going to be replaced, it is often advisable to install coupons of several different alloys to check the performance of each. This allows the engineer to verify his material selection of the new pipe or place on order the material best suited for the particular environment with which he may be dealing.

Coupons should always be used to both verify the probe readings and for long-term verification of corrosion rates.

The advantage of coupons is that they are relatively inexpensive, the weight loss result is positive, samples of corrosion or bacterial product can be obtained from the surface for further observation and testing, and coupons are not subject to instrument failure like the electronic methods of corrosion monitoring. The disadvantage of corrosion coupons is that the results take a long time to obtain and coupons can only give average readings. Coupons will tell you that from when the coupon was installed until when it was removed the corrosion averages so much. Coupons do not take into account that corrosion is not usually constant.

ER probes and instruments

ER probes and their associated instruments measure the resistance through a sensing element that is exposed to the environment to be studied.

The principle of ER probes and instruments is that ER of the sensor, having a fixed mass and shape, will vary according to its cross-sectional area. As corrosion and/or erosion occurs, the cross-sectional area of the element is reduced, thus changing the resistance reading. This change in resistance is compared to the resistance of a check element that is not exposed to weight loss corrosion or erosion and, if the two resistance readings are expressed as a ratio, then changes in this ratio can be shown as a corrosion rate.

Probes are constructed in various designs and materials depending upon the pressure, temperature, velocity, and other process parameters of the system to be monitored.

Instrument availability falls into many categories, but the general configurations of most ER instruments are as follows:

- Portable: portable instruments allow the operator to take manual readings in the field. Newer portable instruments also have the ability to record in the memory facilities the measurements made, tag numbers, and often other data as well. In addition, readings can be observed in the field as they are taken, or they can be left to be uploaded into a PC for recording or charting at a later time.
- Data collection instruments: data collection units can be used anywhere and are especially valuable for use on unmanned or remote sites and locations with difficult access. Measurements are made automatically at the probe, as often as required, and this data is stored onboard for later retrieval, when it is either economically feasible or convenient for the operator to do so. Some data collection units can be programmed in the field or from a PC in the office.
- Transmitter base instruments: in these cases the signal from the probe is transmitted to either a standalone instrument or a PC or incorporated into a full operational SCADA or instrument package for either a full time, scanned time or alarm function readout. The only limitations are the instrumentation package available in each facility.

ER monitoring can be used in almost any environment including a 'dry' system. ER monitoring will also measure erosion. The measurements one gets when using ER technology are averages over time, like coupons, and these readings average the corrosion rate between readings. However, it should be pointed out that the newer instrumentation allows so many readings to be taken in such short times that the averaging time is cut to almost nothing, thus approaching real-time corrosion monitoring. The only drawback to this technique is in systems where conductive depositions are being formed, the deposition may interfere with the resistance readings. Also, as with all sensors, they only measure corrosion when the sensing element corrodes, thus they have a life expectancy that is corrosion dependent. Simply put, probes must be periodically replaced.

LPR probes and instruments

LPR probes and instruments measure the ratio of voltage to current, the polarization resistance, by applying a small voltage, usually between 10 and 30 millivolts, to a corroding metal electrode and measuring the corrosion current flowing between the anodic and cathodic half cells or electrodes. The polarization resistance is inversely proportional to the corrosion rate.

LPR probes also come in a wide variety of configurations and materials. Two and three electrode probes are available, the principal difference being that the three electrode probe attempts to minimize the solution resistance by introducing the third electrode, or reference electrode, adjacent to the test electrode, in order to monitor potential in the solution with a view to reducing the relative contribution of the solution resistance to the series resistance path.

The sensing elements, or electrodes, are made from any material and come in many configurations, usually rod shaped electrodes or flush electrodes. The same considerations in choosing the material must be made as in selecting coupons or ER probes.

Probes are constructed in various designs and materials depending upon the pressure, temperature, velocity, and other process parameters of the system to be monitored.

Instruments are available in portable, data logger or rack-mounted design depending upon the budget and technical need. As LPR gives a real-time result which is continuous, it is much more important to use instrumentation connected to the probe full time, as readings will vary according to the corrosivity of the liquid measured. LPR probes require submersion into an electrolyte and will not function in 'dry' environments.

LPR measurements are much more predominantly used in waterside corrosion environments where fast readings are required, so that chemical corrosion inhibitor being injected can be tuned to prevent the corrosion from taking place or at least kept to a minimum.

Galvanic probes and instruments

Galvanic probes and their related instrumentation expose two dissimilar elements into an electrolyte. These elements of electrodes are attached through an ammeter and the resulting readings offer insight into the corrosion potential of the fluid. It should be noted that worldwide galvanic probe corrosion monitoring is not the most widely used and, although these techniques have their strong proponents, many do not use this technique.

Very much like a coupon, galvanic probes come in many different configurations and materials. The most common configuration of the sensing element is rod type electrodes. Flush elements also are available in several different configurations. Element choice is dependent upon the service and the expected degree of corrosion. Generally speaking, the more sensitive the sensing element, the shorter the life, the less sensitive the sensing element, the longer the life. Depending upon the service and the expected corrosion rates, the proper element shape and sensitivity can be specified.

The sensing elements of galvanic probes are available in any metal material and the same consideration in choosing a material for a coupon does not have to be considered when choosing the material for the galvanic probe sensors or electrodes. Generally speaking the two electrodes must be of dissimilar material. This dissimilarity produces a natural current flow through the electrolyte. Measuring this current flow gives the data to be studied.

One of the more frequent uses of a galvanic probe is for oxygen detection. In this case a brass cathode electrode and a steel anodic electrode are used and changes in the level of dissolved oxygen in treated water will produce readings. Often galvanic probes are installed downstream of water pumps and, if the seals start to leak in the pumps, air (oxygen) will be sucked into the system and the probe will alarm this condition.

Hydrogen permeation technology

Many advances have been made from the original concept of hydrogen probes first used in the 1930s and they will be covered in this section. The two basic categories of hydrogen permeation monitoring devices are those which use a sensor to measure the H_2 resulting from atomic hydrogen permeation and those which actually measure the H_2 resulting from atomic hydrogen permeation through the wall of the pipe, vessel or tank.

Hydrogen permeation, or hydrogen flux permeation, usually originates from the atomic hydrogen atoms that are liberated at the cathode during the electrochemical process of corrosion. It is also possible that the atomic hydrogen atoms may exist due to a chemical reaction inside the vessel or pipe unrelated to corrosion processes. A third possibility is that hydrogen atoms can exist in the steel due to the manufacturing and/or welding procedures. Hydrogen 'bake-out' procedures used today can remove the hydrogen that is present during manufacturing or welding and these procedures should be used. All hydrogen permeation devices use hydrogen permeation, or hydrogen flux permeation, and the resultant H_2 accumulation as the method of monitoring the rate of this permeation.

Not all of the hydrogen atoms generated by a corrosion reaction will necessarily migrate into the steel of the pipe, vessel or tank. Depending upon the process, and the corrosion environment, a certain percentage of these atoms can recombine on the inside surface of the pipe to form molecular hydrogen gas (H_2). This hydrogen gas ultimately goes into the product steam and is lost in the process flow as it is carried down the line. Poisons in the system play a very large role in how much of the atomic hydrogen that is liberated actually goes into the steel itself.

Atomic hydrogen migrates along grain boundaries and therefore not all atomic hydrogen goes in a straight line to the outside of the pipe or vessel. There are theories that some hydrogen atoms can migrate along the wall of a pipe before exiting to the outside wall but, regardless of this theory, we can assume the majority of the atomic hydrogen goes relatively straight through the wall. Once through the wall of either the sensor, the pipe, the vessel or the tank, one hydrogen atom is naturally attracted to another, forming H_2 which either forms inside the annulus of the sensor, escapes into the patch type device or goes off into the air.

Insert type hydrogen probes expose a corroding element to the environment and, as atomic hydrogen permeates the steel and reaches a cavity, the hydrogen atoms combine to form H_2 which builds up pressure in the cavity and this pressure build-up is the basis for the data that may be, and probably is, related to internal corrosion. External H_2 pressure patches also use this same principle.

Another type of hydrogen patch probe is the electrochemical type. This is available as either a permanent patch or a small temporary designed patch that gives a quick reading. This type of device works by polarizing a palladium foil held to the wall of the metal by a transfer medium such as wax, and as the palladium foil is polarized, it acts as a working electrode quantitatively oxidizing the hydrogen as it emerges from the wax. The current induced by the instrument is directly equivalent to the hydrogen penetration rate.

A third type of hydrogen patch monitoring device uses a thin plate designed to capture the molecular hydrogen as it is generated from the atomic hydrogen as it escapes to the outside surface of the pipeline and into the annulus between the outside of the pipe and the underside of the foil.

Since the definable chamber of this H_2 plate is under vacuum, these hydrogen atoms react almost instantly to form molecular hydrogen gas (H_2) once they enter the annulus. H_2 molecules are many times larger than the single hydrogen atom, and are unable to escape from the vacuum chamber by either going back into the pipe or through the foil itself. As more and more atomic hydrogen escapes into the annulus and the build-up of H_2 takes place, the vacuum will then begin to decay or decline in a ratio proportional to the intensity of the atomic hydrogen permeation taking place from the inside surface of the pipe. Use of a vacuum is very important, as conventional pressure build-up hydrogen probes are very susceptible to temperature changes, whereas a vacuum or partial vacuum is relatively immune to temperature changes. Therefore the data is much more even and is not subject to the wide swings that pressure build-up devices experience due to these swings.

Regardless of the cause of atomic hydrogen permeation, the result of this permeation can cause the following to occur:

1. Hydrogen blisters
2. Hydrogen induced cracking (HIC)
3. Hydrogen embrittlement
4. Sulfide stress cracking
5. Carbide phase attack

None of the above conditions are desirable in pressure containing devices that may contain volatile or polluting fluids or gases. In some refinery and petrochemical situations, hydrogen flux flow is extremely high and no appreciable weight loss corrosion is detectable. However, regardless of the source of free atomic hydrogen migrating along the grain boundaries, this flow of atomic hydrogen can cause long-term and/or catastrophic failures.

The quantity and intensity of hydrogen flux are determined by the following factors:

1. Amount of H_2S in the system
2. System poisons (cathodic reaction poisons)
3. Type of corrosion process occurring
4. Type of material of the pipe or vessel
5. Method of material construction
6. Deposition composition

Generally speaking, if the environment in the system has amounts of H_2S, or if it contains cathodic poisons, more of the atomic hydrogen being generated at the cathode will be driven through the wall of the pipe. Although the intensity of the corrosion reaction may vary from system to system, changes in intensity within a given system can usually be directly correlated to the corrosion process. In other words, if a hydrogen flux rate causing a decay in vacuum of 10 kPa/day is reduced to 5 kPa/day, the corrosion on the inside may be cut in half if all other conditions remain the same.

It is this corrosion intensity relationship that allows the user to experiment with inhibitor corrosion control programs or process changes and to be able to see within minutes whether these changes are beneficial, or whether additional changes are required to stop or at least lower corrosion. Similarly, changes in operation parameters can also be observed rapidly.

This newly improved method of near real-time monitoring does not conflict with any existing internal corrosion monitoring devices, but rather adds a new dimension. For example, there is nothing wrong with knowing that there is a general corrosiveness to water and that this corrosiveness increases and decreases. However, the point is, does this manifest itself in actual system corrosion, or are the system corrosion processes driven by factors largely independent of water chemistry?

In summary we can conclude that the development of the many different corrosion monitoring devices and their modification and improvement over time has allowed the pipeline industry to be able to monitor accurately and rapidly almost all of the metal pipes and structures desired. The speed and control of the newest generation of these different techniques has been found to exceed all expectations, with changes in corrosion intensity being measured in minutes and hours, rather than what the previous method offered in days and weeks or even months.

The recent addition of automated systems allowing the data to be transmitted to dataloggers, PCs or SCADA type systems allows for the most modern up-to-date applications. By being able to assimilate this data in a very quick manner, the corrosion events can be compared to other operational parameters allowing the operator to fine-tune these other parameters and thus lower or stop the corrosion.

Advantages of internal corrosion monitoring

The advantages of internal corrosion monitoring include the following:

1. Evaluation and fine-tuning of inhibitor programs
2. Evaluation and scheduling of pigging and other inspection programs
3. Evaluation of process changes and upsets
4. Optimization of the corrosion resistance of an operation
5. Life prediction study verifications
6. Insurance rate adjustment possibilities
7. Management and operation awareness of corrosion or success in corrosion mitigation.
8. Compliance with federal, state, local, and industry rules, practices, and guidelines.

PROTECTION FROM EXTERNAL CORROSION

Protection from external corrosion control involves either putting a barrier between a corrosive media and structure material or by making the structure material a cathode in relation to an anode. The first objective is achieved by providing suitable coating and the second method involves making the structure metal an electrode, where a reduction reaction takes place. We will briefly discuss both methods.

CP

Corrosion occurs where (positive) current discharges from metal to electrolyte at an anode. The objective of the CP is to force the surface of the structure to act as a cathode. That means the current will enter at these spots onto the structure, instead of leaving from the spots.

Direct current flows from the anodic spots into the soil and travels through the soil (electrolyte) onto the cathodic areas. Designing a CP system involves various steps. A typical design flow diagram is shown in Figure 3-6-2. The diagram shows both a galvanic (sacrificial anode) and an impressed current CP system.

There are two ways a structure is made positive: either by use of a sacrificial anode where current is generated due to the potential (galvanic) differences between the anode material and the structure material. The anode material is active and corrodes while structure is cathodic in relation to the anode and is protected from corrosion. Anodes are made out of aluminum, zinc, magnesium or some closely controlled alloy that is most suitable for a given environment

In the impressed current system an external source is used to impress a positive current on the structure to make it cathodic in the soil. The ground bend anodes are not depended upon as a source of electrical energy. Instead, an external source of direct current power is connected between the structure to be protected and the ground bed.

The power sources are various and selected on the basis of their availability and suitability to the conditions and cost.

There are number of techniques available to verify that the system is working:

1. Measurement of structure to environment potential.
2. Test coupons and weight loss measurement.

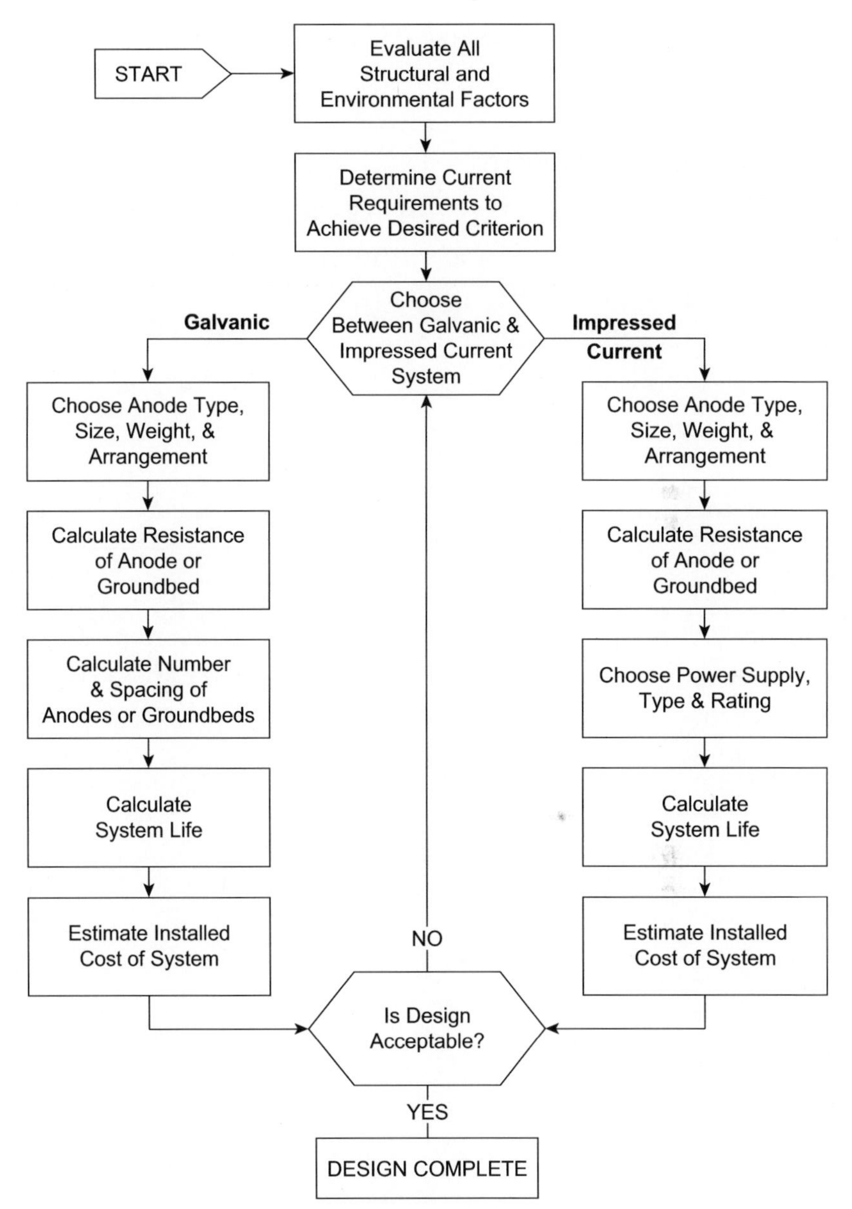

Figure 3-6-2 Typical CP design route.

3. Change in potential between when the current is applied and when current is not applied.

CP protection criteria

The metal-to-electrolyte potential at which the corrosion rate is less than 0.01 mm per year is the protection potential, $E\text{p}$. This corrosion rate is sufficiently low and can be considered an acceptable rate of corrosion for the design life. The criterion for CP is therefore:

$$E \leq E\text{p}$$

The protection potential of a metal depends on the corrosive environment (electrolyte) and on the type of metal used.

The protection potential criterion applies at the metal/electrolyte interface that is the potential which is free from the IR drop in the corrosive environment (IR-free potential/polarized potential).

Some metals can be subject to hydrogen embrittlement at very negative potentials and coating damage can also increase at very negative potentials. For such metals, the potential should not be more negative than a limiting critical potential $E\text{l}$. In such cases, the criterion for CP should be as follows:

$$E\text{l} \leq E \leq E\text{p}$$

A well designed CP system should be capable of polarizing all parts of the buried pipeline to potentials more negative than

−850 mV referenced to a copper/copper sulfate electrode (CSE), and of maintaining such potentials throughout the design life of the pipeline. These potentials are those which exist at the metal-to-environment interface, i.e., the polarized potentials.

For high strength steels with specified minimum yield strength (SMYS) exceeding 80 ksi (550 MPa) and corrosion-resistant alloys such as martensitic and duplex stainless steels, the limiting critical potential is determined with respect to the detrimental effects in the material due to hydrogen formation at the metal surface. Stainless steels and other corrosion-resistant alloys generally need protection potentials more positive than −850 mV referenced to CSE; however, for most practical applications this value can be used.

For pipelines operating in anaerobic soils and where there is known, or suspected, significant presence of sulfate-reducing bacteria (SRB) and/or other bacteria with detrimental effects on pipeline steels, potentials more negative than −950 mV referenced to CSE is used to control external corrosion.

For pipelines operating in soils with very high resistivity, a protection potential more positive than −850 mV referenced to CSE is often considered, for example the following:

$$-750 \text{ m V for } 100 < \rho < 1000;$$

$$-650 \text{ m V for } \rho \geq 1000$$

Where, ρ is the soil resistivity, expressed in ohm-meters.

The protection criteria are discussed in NACE RP 0169 surface to electrolyte potentials in the CP of underground or submerged metallic piping systems. Similar criteria are established in the ISO 15589 and DNV B 401 specifications.

If these criteria are met for steel and cast iron structures, then it is safe to consider them protected by the designed and installed CP system. The following are three CP protection criteria:

1st Criterion

Potential at least as negative as −850 mV Vs $Cu/CuSO_4$ electrode potential, with current applied. Voltage drops other than those across the structure-to-electrolyte boundary must be considered for valid interpretation of voltage measurement. These voltage drops are usually larger in impressed current systems than in sacrificial anode systems.

2nd Criterion

A polarized potential at least as negative as −850 mV. The polarized potential is the same as described in the first criterion except that voltage drops other than those across the structure-to-electrolyte boundary are removed. The most common method is to interrupt the protective current by disconnecting the sacrificial anodes or turning the rectifier. The voltage reading should be taken as soon as possible on interrupting the protective current.

3rd Criterion

This is also called the 100 mV shift criteria, where a 100 mV CP change is measured by interrupting the protective current and measuring the decay of potential with time, or by applying the current and measuring the resulting change in potential of the system.

The application of the 100 mV polarization criterion should be avoided at higher operating temperatures, in SRB-containing soils, or with interference currents, equalizing currents, and telluric currents.

These conditions should be identified and considered during the design stage and prior to using them as test criteria. Furthermore, the criteria should not be used in cases where pipelines are connected to, or consist of, mixed metal components.

Dissimilar metal couples

For systems comprising more than one metal that have different criteria, the criterion for the most anodic material applies to the entire system.

CAUTION ON USE OF STANDARD POLARIZATION POTENTIALS

Under certain conditions, pipelines suffer from high-pH SCC in the potential range −650 mV to −750 mV, and this must be considered when using protective potentials more positive than −850 mV.

Care should be exercised in the use of all protection criteria where the pipeline is electrically continuous with components manufactured from metals more noble than carbon steel, such as copper earthing systems.

For pipelines operating at temperatures above 40°C, the above values may not provide adequate protection potential. In these cases, alternative criteria will need to be verified and applied.

Overprotection by CP

The criteria of a minimum of −850 mV negative is considered to be sufficient to protect a steel structure. The word 'considered' refers to the application of sound engineering judgment. Such engineering determination may include the following:

1. Review of the historical performance of the CP system.
2. Evaluation of the physical and electrical characteristics of the pipe and its environment.
3. Physical verification of the existence of corrosion.

In some situations more negative potential than −850 mV may be required. Examples of such conditions may be the presence of bacteria, elevated temperatures, acid environments, and dissimilar metals. However, more negative potentials can pose some serious problems and that should be one of the points requiring engineering determination.

More negative polarization is termed as overprotection. The possible damages caused by overprotection include the following:

- Coating disbondment may be one of the causes of damage to the pipe system. The cathodic disbonding of coating can happen on any pipe or structure if the coating system's bonding limit is exceeded. Cathodic disbonding of coating is an alkaline phenomenon, causing generation of hydroxyl (OH^-) where the polar bond of coating and substrate is destroyed and the coating does not adhere to the substrate. It may be associated with pre-existing coating holidays or damaged spots, while blisters are formed where the coating did not have a holiday, but lost the adhesion due to high polarization current and/or rising temperature.

- Organic coating systems are more often used to protect buried pipeline. Organic coatings adhere to metallic substrate due to both the mechanical and polar bonding. Mechanical bonding is a result of the anchoring profile achieved by surface preparation methods. The irregular surface profile allows the mechanical bonding. The polar bonding is due to the electrical attraction between polar molecules within the coating formulations and metallic surface where, due to the production of hydroxyl, the polar bond between the coating and the substrate is destroyed.
- CP systems are associated with the evolution of hydrogen. The acceptance criteria discussed above are the minimum potential that will ensure protection of the structure. However the protection level is not uniform and can vary from one section of structure to another. Polarized potential readings of −1015 mV are often recorded. Such high levels of polarized potential signify greater evolution of nascent hydrogen (atomic hydrogen) that can permeate in the material. This is not a major concern for ductile material, however, if the material is stainless steel, titanium, aluminum alloys, pre-stressed

concrete pipe or steel pipe that has higher tensile strength, then more negative polarization may be a problem. The higher tensile strength results in higher resistance to change than the stresses caused by ingress of atomic hydrogen, which can exceed the yield strength, resulting in failure.

Certain materials including steel materials that have higher tensile strength must not be overprotected. Engineering evaluation of the situation is required. The CP design must take this factor into account when adjusting the rectifier output.

Corrosion protection by coating

Coating is applied to a structure to work as a barrier between the corrosive environment and the structure. A well-coated structure requires less current than a bare or poorly coated structure to protect it by the CP method.

To be effective a coating system should have the following properties:

- low moisture permeability
- resistant to chemicals
- dielectric properties
- good adhesion to substrate
- cohesive strength
- good tensile strength
- good flexibility and elongation
- resistance to impact
- able to resist abrasion
- resistance to temperature

- Ease of application and repair
- resistance to cold flow
- resistance to cathodic disbondment

The selection of a coating system is somewhat similar to the selection of material; many factors need to be considered including the following:

a. Cost of coating
b. Type of exposure
c. Surface preparation
d. Operating conditions
e. Any possible upset conditions
f. What substrate is to be applied to its properties
g. Ambient conditions during application (temperature, moisture, relative humidity etc)
h. Environmental regulations, volatile solid particles in the coating system etc.
i. Shop or field application
j. Design and fabrication considerations – imperfect welds, weld spatters, skip welds, rough weld profiles, laminations, gouges, sharp corners, increased areas, gap fasteners, angles, threaded areas, dissimilar metals.

Surface preparation

The majority of coating failures occur due to faulty surface preparation. Several factors that are related to surface preparation affect the life of coatings. Some of these are listed below:

a. Residues of oil, grease, and soil, which weaken the adhesion or mechanical bonding of coating to substrate.

b. Residues of chemicals on the substrate that can induce corrosion.
c. Rust and scale that cannot maintain adhesion.
d. Anchor pattern so rough that adhesion of coating is compromised.
e. Sharp edges, burrs, and cuts caused by cleaning equipment, causing uneven thicknesses on the surface leading to failure of coating.
f. Moisture on the surface if coated over can cause blistering of coating, and delamination may occur.
g. Old coating may be too damaged for recoating to be successful; it may have to be removed to bare substrate and freshly coated.

These conditions can be remedied by surface predation as recommended by the coating system designers. Various industry standards are available to address the needs of a selected coating type. Many tools, techniques and methods can be used to prepare various surface conditions. The table below describes a few of these industry standards.

We have briefly discussed the importance of coating adherence while discussing cathodic disbanding of coating. The importance of surface preparation assumes prime place in the application of good coating on a micro-level and in the integrity management of pipeline systems on a macro-scale. A good coating system must have a well defined surface preparations procedure, selection of application method, monitoring and control of the environment, and inspection and controls to achieve the desired anchor profile and surface roughness, free from contaminates.

Table 3-6-1 Surface Preparation Specifications and Terminologies.

No:	Specification Number	Description
		Following are Joint NACE and SSPC specifications
1	SSPC-SP 5	White Metal Blast Cleaning
2	SSPC-SP 10	Near-White Metal Blast Cleaning
3	SSPC-SP 6	Commercial Blast Cleaning
4	SSPC-SP 7	Brush-Off Blast Cleaning
5	SSPC-SP 12	Surface Preparation of Steel and Other Hard Materials by High- and Ultrahigh-Pressure Water Jetting Prior to Recoating
6	SSPC-SP 13	Surface Preparation of Concrete
7	SSPC-SP 14	Industrial Blast Cleaning
		SSPC Specification
8	SSPC-SP 1-82	Solvent Cleaning
9	SSPC-SP 2-82	Hand-Tool Cleaning
10	SSPC-SP 3-82	Power-Tool Cleaning
11	SSPC-SP 8-82	Pickling
		ISO Standards: Four Grades of Abrasive Cleaning
12	Sa 1	Brush-Off Blast
13	Sa 2	Commercial
14	Sa 2 ½	Near-White Metal
15	Sa 3	White Metal
		Two Grades of Power-Tool Cleaning
16	St 2	Thorough
17	St 3	Very Thorough
		Four Surface Conditions
18	A	Adherent Mill Scale
19	B	Rusting Mill Scale
20	C	Rusted
21	D	Pitted and Rusted

Surface preparation is done in various ways. NACE and the Society for Protection Coating (SSPC) have developed various standards that apply to surface preparation for coating.

There are various more specific application standards like CSA, ISO, and Swedish standards, and DNV standards that must be referred to in specific applications. A list of some of these specifications is given in Table 3-6-1 below.

Index

Note: Page numbers with "*b*" denote boxes; "*f*" figures; "*t*" tables.

C

F

G

H

I

P

W

Printed and bound by CPI Group (UK) Ltd, Croydon, CR0 4YY

13/05/2025

01869691-0001